Agricultural Ethics

Agricultural Ethics

Research, Teaching, and Public Policy

Paul B. Thompson

Iowa State University Press / Ames

Paul B. Thompson is Joyce and Edward E. Brewer Professor of Applied Ethics at Purdue University. He was formerly director of the Center for Biotechnology Policy and Ethics at Texas A&M University.

♾ Printed on acid-free paper in the United States of America

First edition, 1998

The author and publisher gratefully acknowledge permission to reprint the following: INTRODUCTION, "Values and Food Production," *The Journal of Agricultural Ethics,* 2 (1989): 209–23; CHAPTER 1, "Ethics in Agricultural Research," *The Journal of Agricultural Ethics,* 1 (1988): 11–20; CHAPTER 3, "Technological Values in the Applied Science Laboratory," in *New Directions in the Philosophy of Technology,* J. Pitt. ed. (Dordrecht, The Netherlands: Kluwer Academic Publishers, 1995), 139–151; CHAPTER 4, "Emphasizing the Humanities and Social Sciences," in *Agriculture and the Undergraduate* (Washington, D.C.: National Academy Press, 1992) © 1992 by the National Academy of Sciences; CHAPTER 6, "Agricultural Ethics: Content and Methods," *The Peabody Journal of Education,* 67, no. 4 (1990):131–153; CHAPTER 7, "Analyzing Public Policy: The Case of Food Labels," *Journal of Food Distribution Research,* (Feb. 1993): 12–22; CHAPTER 9, "Ethical Issues and BST," in *Bovine Somatropin and Emerging Issues: An Assessment,* Milton C. Hallberg, ed. (1992), © 1992, Westview Press, Boulder, Colo.; CHAPTER 10, "Animals in the Agrarian Ideal," *Journal of Agricultural and Environmental Ethics,* special supplement 1 (Dordrecht, The Netherlands: Kluwer Academic Publishers, 1993): 36-49; CHAPTER 11, "Constitutional Values and the Costs of American Food," in *Understanding the True Costs of Food: Considerations for a Sustainable Food System* (Greenbelt, Md.: Henry A. Wallace Institute for Sustainable Agriculture, 1991); CONCLUSION, "Markets, Moral Economy and the Ethics of Sustainable Agriculture," in *Rural Reconstruction in a Market Economy,* W. Heijman, H. Hetsen, and J. Frouws, eds. (Mansholt Studies 5, Wageningen Agricultural University, The Netherlands, 1996), 39–54.

International Standard Book Number: 0-8138-2806-6

Library of Congress Cataloging-in-Publication Data

Thompson, Paul B.
Agricultural ethics: research, teaching, and public policy / Paul B. Thompson.—1st ed.
p. cm.
Includes bibliographical references and index.
ISBN 0-8138-2806-6 (alk. paper)
1. Agriculture—Moral and ethical aspects. I. Title.
BJ52.5.T539 1998
174'.963—dc21 97–45232

Last digit is the print number: 9 8 7 6 5 4 3 2 1

To

Glenn L. Johnson

mentor and friend

Contents

Preface

Since the chapters of this book were written over a period of fifteen years and for thirteen distinct audiences, a little unevenness in style and presentation is inevitable. In some chapters (like the Introduction) pointers to a broader literature proliferate; in others even direct citations have been kept to the bare minimum. I have changed my mind on several points, too, though I have endeavored to revise where my change of mind represents a major shift of position or point of view. I would like to thank the reviewers and editorial staff at Iowa State University Press for helping me reduce the degree of unevenness, but some amount was inevitable. As noted, each chapter was written for specific audiences on special occasions. In this Preface, I shall attempt to trace their occasional history.

The main part of the Introduction was written to serve as an overall introduction to "values and food production" for one of the working sessions at the Ethics in an Age of Pervasive Technology International Conference, University of Guelph, Ontario, October 15, 1989. Participation was open to the public, and the working session at which I was to present erupted into acrimonious debate immediately after J. Baird Callott's talk on ecological agriculture. I wound up scrapping the talk that appears in this book and substituting some remarks on the ethics of risk assessment and communication. The text did eventually appear as "Values and Food Production," *The Journal of Agricultural Ethics* 2 (1989): 209-223.

Chapter 1 also appeared first in *The Journal of Agricultural Ethics* 1 (1988a): 11-20. It was written for presentation at the Texas Agricultural Experiment Station (TAES) Staff Conference in July 1986. The annual TAES conference allows people in the agriculture program at Texas A&M, where I was then an assistant professor in philosophy and agricultural economics, to compare notes. The audience was a few agricultural scientists, who complained about the technical complexity of the paper. I was not asked to speak at the Staff Conference again for another ten years. I mention this because the paper has always seemed too simple to me.

Chapter 2 was written for a small conference on Ethics in Agricultural Research, sponsored by both the Arts and Humanities and the

Agriculture Divisions of The Rockefeller Foundation and the U.S. Department of Agriculture, Cooperative State Research Service, in Bellagio, Italy, April 8, 1992. My presentation there summarized the paper that appears as Chapter 1 and offered what is published here as an extension and development of those remarks. The Bellagio session was an important turning point in that a number of prestigious agricultural scientists appeared to be taking ethical questions quite seriously, and in collaboration with philosophers, for the first time in my professional experience. These remarks are being published here for the first time.

My participation in the Bellagio conference grew out of a collaboration with David MacKenzie, a plant pathologist of wide interests who has had a long-term interest in ethics, who was one of the conference organizers. MacKenzie contributed a chapter to a book, *Beyond the Large Farm: Ethics and Research Goals for Agriculture,* that I edited with Bill A. Stout (MacKenzie, 1991). The first draft of Chapter 3 was written for a 1990 American Association for the Advancement of Science symposium in which both MacKenzie and Stout also participated. It was revised and presented at the Society for Philosophy of Technology meeting in Mayaguez, Puerto Rico, in March, 1991, before being revised again and eventually published in *New Directions in the Philosophy of Technology,* J. Pitt, ed. (Dordrecht, The Netherlands: Kluwer Academic Publishers, 1995), pp. 139-151.

Chapter 4 was written for a 1991 National Research Council conference on higher education in agricultural science, and published as "Emphasizing the Humanities and Social Sciences," *Agriculture and the Undergraduate* (National Research Council, Washington, D.C., 1992), 208-219. Chapter 5 was written for this volume and has not been previously published. Chapter 6 was requested by Paul Theobald and coauthored with Douglas Kutach. It was published in *The Peabody Journal of Education,* 67(4) (1990): 131-153.

Chapter 7 was written for presentation at a special session on ethics at the Food Distribution Research Society Annual Meeting, Boston, Mass., Nov. 2, 1992, and published in the Society's journal, *Journal of Food Distribution Research* (February, 1993): 12-22. Chapter 8 went through many lives, first as a presentation to the North Central Region meeting on animal behavior in 1984, then as a seminar to the University of Maryland Department of Animal Science in 1986, and even to the Ohio Dairy Farmers Federation Annual Meeting in January 1990. It circulated as a Center for Biotechnology Policy and Ethics Discussion Paper for several years, but it is being published here for the first time.

Chapter 9 was written for Milton Hallberg's book, *Bovine Soma-*

totropin and Emerging Issues (Boulder, Colo.: Westview Press, 1992). Chapter 10 was written for the International Conference on Farm Animal Welfare: Ethical, Scientific and Technological Perspectives, Aspen Institute for Humanistic Studies, Queenstown, Md., June 7-10, 1991, and published in the *Journal of Agricultural and Environmental Ethics* 6, special supplement 1 (1993): 36-49. Chapter 11 is a substantially rewritten version of a talk given for the Wallace Institute for Alternative Agriculture in 1991 and published as "Constitutional Values and the Costs of American Food," in *Understanding the True Costs of Food: Considerations for a Sustainable Food System* (Greenbelt, Md.: Henry A. Wallace Institute for Sustainable Agriculture, 1991): 64-74.

The Conclusion was prepared for the Inaugural Conference of the Mansholt Institute at Wageningen Agricultural University, The Netherlands, and published in *Rural Reconstruction in a Market Economy*, W. Heijman, H. Hetsen and J. Frouws, eds. (Mansholt Studies 5, Mansholt Institute, Wageningen, The Netherlands, 1996), 39-54.

All of the republished materials are used with permission.

Acknowledgments

The number of readers, editors, research aides, and secretaries who have helped at various stages in the writing of this book is beyond my ability to recall. I would like to thank Daralyn Wallace of Texas A&M, who was primarily responsible for helping me with the final manuscript preparation, and the editors and readers for Iowa State University Press. I would also like to acknowledge the editors and journals noted in the Introduction who have agreed to the republication of these materials.

Research for various parts of this book was supported by two National Science Foundation Grants from the Program on Ethics and Values Studies. I wish to thank Rachelle Hollander at NSF.

All of the essays in this book were written during my tenure at Texas A&M University. I extend heartfelt appreciation to my students and colleagues at Texas A&M and to all the citizens of the great state of Texas.

Agricultural Ethics

Introduction

Values and Food Production

This book on agricultural ethics may appear to come from out of the blue, without advance notice, and without the literature and reference points that are characteristic of most philosophical topics. So far, there has been little continuous research and publication on the ethical choices in food production in our time, though the growth of such work in the past decade has been remarkable. It is, therefore, unlikely that we bring a common vocabulary, let alone a common agenda, to the discussion of ethical choices for food systems. This book aims to provide an introduction to philosophical reflection on agriculture and food production by using some key concepts for ethics within a discussion of food system issues and then defining these concepts in rudimentary terms.

With only a few exceptions, the chapters in this book were written for special occasions where the philosophical approach to ethics was being brought to an audience that could not be expected to have prior familiarity with or orientation to it. My aim was to say something that my audience would immediately understand as making a difference to how one thinks about or evaluates agricultural and food distribution practices. Although I hoped that my audiences would be motivated to think more deeply about these practices, I did not expect them to adopt a comprehensive moral theory on the basis of my remarks. Since philosophers often address each other with the aim of critiquing or defending comprehensive moral theories, and since classroom teaching in philosophy is often intended to familiarize students with these philosophical debates, many professionally trained philosophers have few opportunities to write occasional essays for the general public. As a result, philosophers who speak to audiences preoccupied with more immediate and practical problems often fail to connect. That explains why there has been little writing and discussion on agriculture from the side of the philosophers.

The primary rationale for collecting and reprinting these essays is, thus, not to reach professional philosophers, but to continue and extend the practical purposes for which they were originally written. The book aims to illustrate the relevance of philosophical ethics to practical matters, to equip readers with a first-level capacity to interpret and utilize the basic vocabulary of philosophical ethics, and to stimulate deeper inquiry into the moral and cultural dimensions of agriculture and food systems. The three main sections of the book—research, teaching and public policy—are organized with these ends in mind. Each section begins with two pedagogical essays, whose purpose is to introduce terminology and basic approaches. These essays are then followed by at least one chapter intended to broach a more complex and enduring philosophical issue.

American agriculturists will recognize an analog to the "three-legged stool" of research, teaching, and extension in the title and organization of this book. The three-legged stool summarizes the elements of the land-grant model for agricultural research. American land-grant universities (together with the U.S. Department of Agriculture) were shaped by a philosophy that saw scientific research, formal education, and continuing education as facets of a unified whole. Following the German tradition, science would be applied to agriculture, but America's rural people were to be educated in the principles of science, too. This education would prepare them for lifelong learning offered through the state and county extension services. Endowed by grants of land authorized in 1862 and 1890 (hence their name), American agricultural colleges reflected this philosophy in their organizational structure. Generally (and to a varying extent) a single faculty staffed the agricultural experiment station resident instruction, and extension services in each state. The three legs of the stool were thus integrated in the person of each faculty member in the agricultural program.

Since most of the individuals recruited to these faculties had farm backgrounds, there was also continuity with the values and life experiences of the producers, who were the expected recipients of the wisdom and technology that was being generated by agricultural faculty. The common demographic base of research faculty and farm producers gave them shared values and a common life project, but it did not give them any reason to discuss or debate either their values or their project. As a result, the agricultural ethic of the first century of agricultural science was implicit, and the bearers of this ethic had no incentive to make it explicit. That explains why there has been little discussion of agricultural ethics on the part of the agriculturalists.

Do farmers, scientists and educators need a reason to undertake an explicit discussion of agricultural ethics today? There are at least three reasons to think that they do. First, they clearly cannot rely on implicit shared values derived from common life experiences to shape the future. Tomorrow's agricultural scientists will not have grown up on farms, and neither will the vast majority of political leaders (let alone citizens) who will shape agricultural policy. The communication of agriculture's ethic must be done explicitly. Second, changes in agricultural production *and* social values are reflected in a number of contested political issues on which the farm community does not speak unequivocally. Values, in short, are less shared than they were. Third, farmers, researchers, agribusiness firms, and educators face challenges in the twenty-first century that derive from a century of success in increasing yields, compromising environmental quality and depleting their own ranks. There is thus ample reason to rethink the premises on which twentieth-century science and policy were built.

This book aims to stimulate discussion and to improve the quality of debate, *not* to provide a comprehensive approach to all the ethical questions that might be associated with agriculture and food production. In this Introduction I paint a picture of where and how ethical issues arise. There are three main ways. First, new production technologies produce unintended consequences or involve risks of unwanted consequences. Second, society's grasp on ethics itself changes, just like commerce, engineering, or agricultural science. As a society, we are constantly exploring the next implication and application of fundamental ethical values, and sometimes, as in the abolition of human slavery, this exploration leads us to reevaluate practices that have a tremendous effect on agricultural production systems. Finally, tension occurs when culture moves forward in both science and ethics, leaving old ways behind. The philosophical significance of such tension cuts deeper and is more difficult to articulate than the problems created by unwanted consequences or new moral understandings.

Traditional Values for Agricultural Production

Major philosophical innovations in ethics occurred within a relatively short span of time beginning in roughly 1650 and culminating in the last quarter of the eighteenth century with the writings of the first utilitarians and the ethical rationalism of Immanuel Kant. Although some will object to my speaking of a culmination here, it is fair to say utilitarian thought plus some amalgam of Kantian thought and the rights theory of the French philosophers now dominates secular study of ethics and political theory in the United States. Alfred North Whitehead wrote that all philosophy is a footnote to Plato, but it is more literally true that recent ethical thought is a footnote to Bentham and Kant. Furthermore, the ethical innovations of the European Enlightenment penetrate deeply into the consciousness of everyone who has been brought up and educated in the Western intellectual tradition. The utilitarian conceptions of trade-offs and optimization share a place with the rationalist concepts of liberty and rights in the ethical vocabulary of virtually everyone. Religious and philosophical ideas that predate the Enlightenment continue to have an enormous influence on the average person's perception of ethics and values and we can come to a sharper understanding of how traditional values influence agricultural production if we consider the break between traditionalism and the Enlightenment more carefully.

Enlightenment philosophers were consciously responding to several felt needs in proposing new ideas for ethics and, indeed, for philosophy and natural science, as well. One was certainly a change in the shared perception of ecclesiastical authority, but this theme shall be set aside for our purposes. Fundamental material changes were also taking place in European society. European economies were completing a transition from predominantly rural societies organized under the manorial system of villeins bound to a liege lord to diversified societies organized around manufacturing and trade centers in the cities and towns. These changes produced two phenomena that must have been highly conducive to change in ethics. One was the opportunity for individuals to advance their personal interests as societal institutions changed;[1] the other was a sense of anomie, as people found the role morality of the feudal age increasingly inapplicable to their new situation. The first phenomenon provided an incentive for people to abandon traditional moral codes; the second provided an incentive for replacing them with something new.

Appreciating the code that was being abandoned (albeit in piecemeal fashion) is crucial to appreciating the Enlightenment changes in moral thought as well as for understanding the persistence of traditionalist ideas for our own time. In the manorial system, peasants worked land owned by aristocrats. In addition to being the owners or bosses of the production process, the aristocrats held political rank, and the peasants owed political as well as economic obligations to them. In the manorial system, participating in primary agricultural production activities was part and parcel of duty to one's lord. The key moral concept was one of personal loyalty, in deed if not in mind. The ethical code of both peasant and aristocrat could have been spelled out almost entirely in terms of loyalty to specific individuals who were well known to those who had duties to perform. In contrast to Enlightenment ethical thought, which stresses the universal—the generally applicable principle—medieval duties were particular, owed to a particular person, and to be performed in a particular place at a particular time. Even the agricultural price system was grounded in traditional fees and tariffs specified in terms of particular obligations owed between miller and peasant, peasant and lord. Any attempt to exploit an advantage arising from supply and demand was regarded as an attempt to shirk a moral and political obligation to exchange at the traditional rate (Moore, 1966; Rosenberg and Birdsell 1986).

It is significant that the manorial system was agricultural. Traditional values for agricultural production are, in fact, traditional values for society as a whole. Any given plot of land is unique, entirely particular. Tying obligations to a particular place was feasible, even natural, for agriculture, because the land itself had a permanent location that provided a measure of security to medieval relationships. As long as the peasant stayed on the land, the lord could not abandon him entirely, for the lord's own interests were in the land, and these interests could not be easily moved. Furthermore, it was land that knitted political and economic relations into a seamless web. Politics of the feudal era were essentially concerned with territory. Political rank was generally based on domain over a specific parcel of land. People who could move their production from one territory to another could change their political loyalties, but agricultural producers could do so only by abandoning their claim to their key productive asset, the land.

The ethic of particularized loyalty, however, became dysfunctional as societies came to emphasize trade and manufacturing. Both of these activities required entrepreneurs to take risks and to deal with much larger circles of people. They did not require that their practitioners remain in the same place indefinitely, and sometimes presented great incentives for relocation. Further, the morality of location-bound loyalties did not support the norms required for trade. As trade and manufacturing spanned the globe in search of materials and markets, the serviceability of a moral code based solely on personal loyalties to particular individuals became obsolete. Commerce required rules for upholding covenants with people from many different places. Trading societies needed an ethic whereby people could be trusted to honor contracts, and it needed a legal system that could enforce them.

The ethical content of the contract was, in contrast to the particular duties of personal loyalty, spelled out by the entirely impersonal rights and privileges specified in the terms of exchange. Contractual agreements did not depend on the uniqueness of their particular signatories, and, indeed, one could agree to the same set of contractual terms with two, a dozen, or a hundred people. Moral duty consisted not in loyalty to a particular person but in fulfilling the specified obligations, without regard to the particular identity of the other signatories (Rosenberg and Birdsell 1986, 115-134). It should not be surprising, therefore, that philosophers got around to thinking of morality in contractual terms, either literally, as in the case of Hobbes, or as a similarly faceless and universalized specification of entitlements, rules, expectations, and obligations in the case of Bentham and Kant.

Now this is an enormously condensed version of a very long story and one that demands much more detail before any interesting historical questions can be raised. Yet we can ignore the question of what caused the arrival of urbanized trading and manufacturing societies, as well as the question of what caused the rise of new morality. The point here is simply to note two kinds of functional fit between morality and the material organization of the economic system. The traditionalist morality of the manor had a logic that would have tended to shape intentions and behavior in ways that could be supported by its dominantly agricultural forms of production. Traditionalist morality was not so well-suited to trade and manufacturing. However, the rising moral ideas of the Enlightenment placed importance on compliance with a fairly abstract specification of rules and expectations. The new morality was suited to relations among relative strangers in a way that traditional morality was not. The implications of the new morality have not been entirely realized, even yet (Moore, 1966).

Changes in Agricultural Production Technology

As Europe became more urbanized, agriculture changed as well. The manorial system changed slowly, but by 1700 it had been substantially displaced by a system of economic organization for agriculture not unlike the one that exists throughout Western Europe, Canada, Australia, and the United States today. Owners of the land produce for profit. Labor is supplied either by the owner, by wages, or through a contractual agreement. There are, of course, land leases and absentee landlords, but in all cases agreements between owners, producers, or hired workers are contractual, subject to negotiation, and influenced by market forces. They do not entail political obligations over and above the respect for property rights. There is, thus, some sense in which agriculture adopted the contractual morality of the towns. At the same time, the family farm, a production unit that came to dominate agriculture in the northern regions of both Europe and North America, retained a large measure of the traditionalist morality of a bygone day, though with the understanding that the owner-operator was to enjoy the same kind of dominion over the land that the aristocrats enjoyed in feudal times. The remnants of traditionalism in agriculture is a theme that recurs throughout this book.

The rise of contractual, Enlightenment morality within Europe and its eventual transfer to the European colonies created a situation that invited technological innovation in agriculture. The manorial system of agricultural production had provided few incentives for adopting new production techniques, and traditionalist morality reinforced this situation by making the repetition of existing cultural practices into a political obligation. Fees and tariffs were set by tradition, and greed was one of the seven deadly sins. When the feudal system of political obligations was shattered, new economic opportunities for agricultural producers emerged as well. At the outset, these opportunities were primarily captured by the large land-owning families that had dominated the feudal era. The enclosure movement broke up the common lands of the manors, converted them to commercial production, and forced many peasants to the cities and towns to seek employment. John Locke's defense of the enclosure movement stated what was to become the dominant ethic for agriculture well into the present day:

> He that incloses land, and has a greater plenty of the conveniences of life from ten acres, than he could have from an hundred left to nature, may truly be said to give ninety acres to mankind: for his labour now supplies him with provisions out of ten acres, which were but-the-product of an hundred lying in common.

In the colonies (and eventually in Europe itself) ownership patterns became more diffuse, but the pattern of producing at least a segment of the crop for profit, rather than self-sufficiency, was imperative in a society in which many people were engaged in trade and manufacturing. The same rationale that Locke gave for enclosure provided moral support for new technologies that increased the productivity of agricultural land and, not incidentally, enriched those who were quick to adopt them.

The succession of technological innovations in the past 100 years includes mechanization, use of fertilizers, new crop varieties and animal breeds, insecticides, herbicides, improved transport and communications, and, now, computers and biotechnology. New technologies that increase productivity and increase the owner's control of the firm become almost universally adopted by all producers of a given commodity, given constraints on the technology's applicability to climatic and geographical conditions. As a given technology increased the aggregate productivity of all producers, unit prices of the commodity tended to fall, passing long-term benefits on to consumers. The ethic for the primary decision maker was simply one of self-interest, but farmers and their suppliers could rest easy with the thought that market forces and increasing productivity extended benefits to everyone in society. It was Locke's logic, refined and developed through the concepts of free enterprise philosophy and utilitarian ethics. Sloth replaced greed in the seven deadly sins.

Humankind does not live by bread alone, however, so it should not be surprising that the single-minded pursuit of increased yields ultimately came into conflict with other values. Some of the new technologies had unwanted side effects. They increased yields, but they increased soil erosion, polluted water supplies, or otherwise exposed people to toxic substances. The moral arithmetic for justifying production techniques with unwanted outcomes is far more complicated than Locke's simple formula, and it is a problem with which economists and philosophers have struggled mightily during the last two decades. It is no mystery, however, why there is a problem here. New and bigger technologies have increased the risks of agricultural production, while at the same time new biomedical technology has made us more cognizant of risks that we may have run unknowingly in the past. There is a problem because agricultural technology exposes people to risks, and the primary and secondary decision makers who develop, employ, and regulate these technologies have not secured the people's consent.

Recall from the discussion earlier that risk and consent are at the heart of the contractual morality of the European Enlightenment. European trades needed a moral code that allowed them to accept risk—something that traditionalism did not do—but simultaneously to control risk. The purpose of the contract was to enable commerce by specifying duties, hence limiting the risk of the contracting parties. At the same time, the trader accepts other risks in hopes that the enterprise will turn a profit. The contractual situation is one in which each party consents to bear certain risks and expects to ensure against other risks for having done so. John Locke's portrayal of the social contract (which serves as the basis for civil authority) stresses the individual's incentives for relinquishing natural freedoms in exchange for an assurance that one's life, liberty, and property will not constantly be at risk from usurpation (Locke 1690, 144). The social contract is a balance of risk and consent. To the extent that contractual thinking exemplifies Enlightenment morality, the failure to secure consent for the imposition of risk goes against the deepest grain of the modern age. Of course one cannot secure consent from every individual before one sprays a field or invents a plow. The ethics of risk and consent cannot be summarized in a paragraph, and there is much more to say about the conditions under which consent is implied, and when representatives of the people are and are not authorized to give consent on their behalf.[2]

The fact that we have three centuries of experience in how to think about these issues, and in how to deal with them politically, does not mean that we always deal with them successfully. The point here is simply that our technology has created ethical problems, but they are problems of a very familiar sort. Agricultural producers and those who support them with technology may have been seduced into thinking that, so long as they increased food availability, they were exempt from the constant process of politically negotiating and renegotiating the moral bargain that is at the foundations of the modern democratic state. They may have felt ambushed by Rachel Carson because in 1962 they really had not had to pay too much attention to the politics of risk. A quarter century later, there can be no such illusion. Democratic societies will not entrust their water, their diets, or their natural resources blindly into the hands of farmers, agribusiness firms, and agricultural scientists. Agricultural producers must participate in the dialogue that leads to social learning and social consensus about risks, and they must be willing to contribute the time and resources needed to understand the positions of their fellow citizens, and to make articulate statements of their own position. This is not a new responsibility, though it may have been neglected in the past.

The chapters on agricultural research begin this theme in the book. They examine how some well-established ethical concepts can be applied in evaluating agricultural technology. The themes of risk and consent are also taken up in chapters 5 and 6 on education, and in Chapter 7, "Analyzing Public Policy: The Case of Food Labels" and Chapter 9, "Ethical Issues and BST." It will be evident that unintended consequences of technology tend to dominate the problems taken up in this book, and that the primary method of approach to them is one of applying standard philosophical concepts. As Chapter 3 on "Technological Values in the Applied Science Laboratory" indicates, this should be viewed as a pragmatic response to particular problems rather than as a consequentialist or utilitarian philosophical orientation. Many issues in contemporary agriculture are seen as problematic in virtue of the unintended consequences of technology. My philosophical practice is to accept an existing "problematization" (e.g., a way of seeing a given issue as problematic) unless I see compelling reasons to "reproblematize" the issue. Chapters 9 on BST and 11 on constitutional values involve some modest reformations of issues, but I do not reformulate issues simply for the sake of philosophical display.

The Growing Scope of Morality

If ethical obligations and expectations that govern agricultural production are entirely specified by known loyalties to specific individuals, they are somewhat limited and one would not expect substantial change in them over time. The attempt to specify moral obligations in more general terms has not been so easily contained. In the case of Bentham and the utilitarians, morality is specified as a general obligation to optimize the balance of pleasure and pain or, to use more contemporary language, the quality of life for the greatest number. Particularizing features such as social rank, sex, race, or creed are irrelevant in this formula; all affected parties are counted in the optimization problem. Bentham himself was aware that even the attempt to exclude non-human animals from the list of affected parties was arbitrary. Utilitarianism, thus, became a conceptual battering ram against the bulwark of traditional rules that excluded first the common classes, then the landless poor, then religious minorities, women, racial minorities, and, most recently, even animals from moral consideration and protection by the state.

Rationalist philosophers preferred to modify the traditional notion of natural rights so that the possession of rights depended not on having received them as a gift from higher authority (be it God or king) but as a consequence of the rational capacities of human beings. One cannot rationally deny to others what one must have for oneself in order to act rationally at all. To do so would be to claim for oneself a kind of privilege that would never be recognized in another. As in utilitarian thought, rationalist thought attempted to describe a moral point of view that had both claim and appeal on all persons, without regard to the particular accidents of birth and, like utilitarian thought, it tended to sanction an expansion of the moral point of view to include commoners, the poor, women, and non-Europeans. It, too, has recently been extended to include non-human animals.

A new group of problems in identifying the values to guide food production can be traced to the tendency of modern ethical philosophies to extend the scope of morality well beyond the guideposts of traditional rules. Our society has already assimilated some of these extensions. Food production is now uncontroversially regarded as production of commodities for sale in markets, rather than as the performance of a political obligation owed by a peasant to a feudal lord. We are struggling with the idea that all the poor might be entitled to a share of food production, but we seem to be moving toward a consensus that, however such rights are assigned, they will not interfere with the current distribution of rights and privileges possessed by farmers and ranchers. This question, thus, appears to have only minimal relevance to the primary and secondary production choices noted earlier.

Of far more current importance are three other ways of extending moral consideration. The first, already mentioned, extends the primary terms of moral theory to non-human animals. The second extends consideration to unborn generations. The third attempts (so far, less successfully) to extend the main concepts of ethical consideration to nature itself. Each of these extensions is surveyed in highly abbreviated form. The strategy is to examine how extending the scope of either utilitarian or Kantian ethical ideas might be expected to influence the primary and secondary types of choice that have been identified.

Non-human Animals

The utilitarian model of ethical choice (discussed at more length in Chapter 1) requires the decision maker to consider all the consequences of a proposed action and to select the proposal that promises the best consequences. Utilitarian theory (sometimes called consequentialist theory) offers many alternatives for defining best or optimal consequences, but the selection of a decision rule is not what concerns us here. The utilitarian decision maker must consider consequences for all affected parties, and although this consideration applies primarily to human beings, non-human animals are affected by our actions, too. When Bentham proposed the principle of utility in 1789, he was thinking of consequences in terms of sensory experience, pleasure or pain, and he was quite aware that the logic of his approach to ethics entailed consideration for the sensory experiences of non-human animals (Bentham 1789). Contemporary utilitarians are more likely to talk about the satisfaction of preferences or of the quality of life than they are pleasure and pain, but the infliction of suffering on any creature must be counted when the calculation of benefit and harm is completed. Peter Singer is the contemporary philosopher who is best noted for defending the welfare of animals on utilitarian grounds (Singer 1975).

Farmers or ranchers who make production decisions based on

whether they produce optimal outcomes for all affected parties will, if this extension of the utilitarian ethic is correct, take animal interests into account. This may lead them to reject some methods of animal production altogether, and it would almost certainly make them inclined to accept some loss of profit to improve the well-being of their animals. Notice that the utilitarian view does not entail total prohibition of humane slaughter, for it is entirely possible that whatever harm is done to the animal in such an instance is fully repaid in benefits to human beings. There is also room for maneuver in calculating both the degree of harm to the animal and the degree of benefit to human beings. Singer and the animal welfarists see great harm done to the animal in modern confinement agriculture and comparatively little benefit to producers and consumers of animal products. One might, however, accept the utilitarian view that animals count, admit that their suffering should be minimized, but deny that the evaluation of consequences demands any significant departure from current animal production methods.

More radically committed animal activities are likely to take the view that animals have rights that must be respected. As explained by Tom Regan, the idea is that animals have subjective experiences, feelings of pleasure and pain, at least, and probably more complex perceptions of their interests and attachments. Though other mammals have poorly developed mental capacities when compared with humans, they are subjects of a life, just as we are, and this is the characteristic that entitles them to rights. Being the subject of a life is the fundamental prerequisite to purposeful action; failure to respect the subjectivity of others denies the principle of right action. Regan and the animal rights activists understand rights as moral claims on the actions of others, not simply as legal entitlements. Moral rights are only meaningful if they are grounded in a right to life and in basic liberties that give the creature an opportunity to exercise its capacities. As such, the doctrine of animal rights entails prohibition of raising and killing animals for food, though it might permit restricted forms of egg and dairy production (Regan 1983).

The welfare and rights positions for animals, and their implications for human producers are reviewed in Chapter 8. The point here is to disclose the origins of this debate in the continuing growth of ethical ideas that emerged in the European Enlightenment. Because utilitarian and Kantian philosophies construe ethics as impartial and impersonal, the impulse to confine moral consideration to the human species becomes, on reflection, arbitrary. The trading societies of the Enlightenment, as well as those of our own time, needed an impersonal morality. Enlightenment morality is based on having goals, interests, or preferences, rather than on family or class loyalties, and though humans certainly have goals, interests, and preferences, so, probably, do other animals.

Future Generations

In one sense, traditionalist morality was highly considerate of future generations. Because it both presumed and tended to reinforce social and technological stability (some might say stagnation), it provided future generations with virtually the identical opportunities and privileges enjoyed by present ones. Nevertheless, it is fair to say that these opportunities were not provided on behalf of future generations, and the loyalties of the manorial system were not addressed to nonexistent people. The utilitarian and Kantian ethics that have displaced traditionalist morality are often cited in support of social change; they do not promise or endorse stability of the status quo for future generations. At the same time, for the same reasons that Enlightenment ethical systems can be extended to non-human animals, they can be extended to non-existent humans.

For utilitarians, the nature of this extension is quite simple. Some of the consequences of an agent's acts will clearly be endured by individuals as yet unborn. Because all affected parties are to be considered there seems to be no reason to think that a person in a future generation who will be affected by present acts should be excluded from an evaluation of the consequences of an act. Similarly, one can imagine acts that violate the rights of the unborn, if, for example, a present act would endanger the lives of people in future generations, or would deprive them of important opportunities. Kantian or rights ethics can also be extended to include the rights of future generations. In point of fact, extension of both utilitarian and Kantian notions to future generations has proved to be far more complex than extension to animals,[3] but in the interests of simplicity, discussion of these difficulties will be omitted here.

Concern for future generations has been one key basis for environmental activism throughout the past two decades. People have widely accepted the notion that the present generation is morally obligated to pass an ecologically whole environment on to our progeny. This concern for future generations, also an extension of Enlightenment morality, has far-reaching implications for agricultural producers and for those who design, manufacture, sell, and regulate agricultural production. Farm families have traditionally felt continuity between parent and child, grandparent and grandchild, and so on through a sequence that links the generations. This type of intergenerational concern is consistent with the traditional model of moral obligations established on the basis of loyalties owed to particular persons. It differs, therefore, from the extension of Enlightenment morality to future generations. The traditionalist cares about children and grandchildren that will carry on his or her particular line; duties are owed to particular individuals, who, even though unborn, can be uniquely identified as the descendants of specific present day people. The more general duty to future generations is a generalized duty to the people who will inhabit the earth in the future, whoever they might be.

The most obvious way to incorporate concern for future generations into a vision of sustainable agriculture is to interpret permanent resource depletion and long-term pollution as a cost of present day production that must be born by posterity. Such an approach was implicit in the attempt to assess the true cost of food, that is discussed in Chapter 11. The "true costs" approach has met with immediate resistance from economists who point out that the present day value of costs born in the distant future must be "discounted." The present value of a future cost is equal to the amount that would have to be invested today in order to compensate tomorrow's loser. Since we expect that compound interest would be paid on that investment, the present day value of impacts in the distant future are very small indeed.

In Chapter 11, I argue that both sides in this debate fail to recognize the way in which monetized costs reflect only the shallowest and least enduring of our values. I suggest instead that sustainability must be framed in light of constitutional values. The book's conclusion represents my most recent thinking on the question of sustainability, and it returns to the organizational theme of the land-grant triad by discussing the research, education and lifelong learning dimensions of a commitment to sustainability.

Environmental Extensionism and the Limits of Enlightenment Morality

There are many environmental activists who are uncomfortable with thinking that future human use of resources is the reason for adopting sustainable methods of agricultural production. Speaking in terms of obligations to future generations seems to imply that nature and natural systems are of value because they are or will be useful to human beings. The suggestion that human use is the sole criterion of value for nature has been explicitly rejected by many prominent naturalists and environmental activists (Leopold 1947; Schumacher 1972) and more philosophically inclined environmentalists have attempted to find a way of extending the main categories of Enlightenment ethics to plants, forests, and other ecosystems, or perhaps to nature as a whole.[4]

This attempt at extension has been less successful than the previous two. For utilitarians, value refers to the act of being valued by a sentient being. This makes it fairly obvious how things valued by non-human animals might be included in our deliberations, for a human act that deprives a brute of something it values has morally significant consequences. Asserting that non-sentient beings, such as plants, have preferences or otherwise perform valuative acts is less plausible, and asserting that groups, species or ecosystems take an interest in what happens to them is so far from traditional utilitarian thinking that it cannot be classified as an extension of Enlightenment morality. For Kantians, who rely on rights, duties, and respect as their fundamental moral notions, the extension of Enlightenment morality goes a little further. They must show that non-sentient beings such as plants and ecosystems have interests, that they are entitled to respect, and that, therefore, humans are obligated to act on their behalf in securing those interests (Stone 1974; Taylor 1981).

Opponents of this philosophy have argued that this extension is really an equivocation on the word *interests* (Frey 1983; Attfield 1983), but the details of this debate are less relevant than the fact that having gotten in the habit of extending the scope of normality, we have become accepting of arguments that employ this strategy in novel areas. If Enlightenment morality could be extended to concern for the interests of natural ecosystems and species, the result might well be more severe for agricultural production than concern for sustainability. Agriculture is, after all, the intentional disruption and destruction of ecosystems as they evolved in their natural state. If this third form of extension is valid, perhaps we should regard agriculture as a necessary evil, and we should try to do as little of it as possible. Perhaps we should dive headlong into biotechnologies that would allow us to grow our food in vats so that we could leave the largest part of nature undisturbed.

Giving moral consideration to nature itself seems to reach the limit of extensionism, of extending Enlightenment morality to new interests and to new goods. Radical or deep ecology seems to require not an extension of Enlightenment ideas, but a revolution no less fundamental than that of Enlightenment morality itself. An ecological revolution in values would place morality on a new conceptual foundation, and being new, it is hard to grasp what such a foundation might look like using tired old Enlightenment concepts. A revolution in moral thought may indeed by on the horizon, but this is not the place to take up that subject. What seems more likely to me is a regression to the traditionalist moralities of our feudal past. They have never completely left us for over 400 years, and they continue to be influential in agricultural issues today. Several chapters in this book discuss this "new traditionalism," but the environmental ethics dimension of that discussion was the primary topic of *The Spirit of the Soil* (Thompson, 1995), and the central claims of that book are not repeated here.

The New Traditionalism

Thus far we have identified two types of change and seen how each gives rise to moral concern about agricultural production. The first type of change is changing technology. New technologies produce unintended consequences, and our attempt to evaluate these unintended and uncertain consequences brings moral considerations to bear on production decisions in new and unsettling ways. Questions about food safety and environmental quality loom large in this category. We have also experienced a second type of change, however, in the application of morality itself. Extension of moral concern to non-human animals has raised questions about farm animal well-being and animal rights. Extension of moral concern to future generations has raised questions about the sustainability of agricultural production. Extension of moral concern to plant and animal species and to natural systems provides the basis for a radical environmentalism that portrays agriculture in a darkly unfavorable light.

This survey covers many of the value issues that commonly appear in ethical reflections on agricultural production. It omits some issues that are of vital importance simply because they are more frequently related to agricultural distribution and consumption—world hunger and population issues, for example. Those who feel that our technical capability entails a responsibility to solve distribution and consumption problems that have been with us since the dawn of civilization may want to include these issues under the category of technological change. It is not likely that such problems will be resolved by innovative production technology. They address a different class of value concerns altogether and these, too, have been omitted from the volume. Readers should consult William Aiken and Hugh La Follette, *World Hunger and Moral Obligations,* 2nd edition. (Englewood Cliffs, N.J.: 1995, Prentice-Hall.) I have written on the philosophical debate over world hunger in *The Ethics of Aid and Trade* (Cambridge and New York: 1992, Cambridge University Press.)

Even excusing this omission, however, an approach aimed only at considering technology's consequences and the extension of moral concepts fails to touch on one question that has been prominent in every U.S. production policy debate since the turn of the century and has analogues in most industrialized nations around the world. What is the value of the family farm? Is there a moral obligation to save family farms? One might think that this question belongs in the category of technological change. It is common knowledge that changes in production technology create several trends that militate against relatively small family farms. Technological innovation changes production efficiencies; this in turn changes economies of scale and, more important, creates the treadmill effect whereby farmers who innovate run faster to stay in the same place, while those who fail to innovate fail to survive. If small family farmers are technologically conservative (e.g., reluctant to adopt new technology) the treadmill effect constitutes a bias against them. Even when they are not conservative, the economic climate in which farm failures are accompanied by windfall profits to innovative farmers may well mean that successful farms grow larger. Technological change in other areas affect small farms, too. For example, transportation and information technology is partly responsible for the large supermarket chains that prefer to contract with large-scale suppliers.

There is no disputing that technological change has made agriculture more competitive, and that this has sometimes made life more difficult for family farmers; but it has made life more difficult for harness makers, too. Simply noting these difficulties falls short of identifying a philosophical problem. Technological impact on the size distribution of farms is not morally significant unless we have some reason to think that the continued existence of family-type farms is valuable in the first place. This is not to say that the harm caused by farm structural change is insignificant. Enlightenment morality provides many reasons to think that harm to economically displaced individuals is very significant, but it is equally significant without regard to the occupation from which the individual is displaced. As such, while we may want to assure that suffering is minimized, or that losers are compensated, or that small farmers' rights are not violated, we have no reason so far to be concerned about small farms as institutions. Even if we talk about the economic health of rural communities, we do not find a basis for moral concern about the demise of family farms understood as a social institution, for a rural community may do just as well with a tire factory or a rendering plant on its outskirts as it does with a few hundred small farms.

Enlightenment morality, however it is configured, aims to protect and advance human interests in universal terms. Although Kantian ethics, for example, can explain why it is important that individuals have a high degree of personal autonomy in choosing and pursuing their careers, Kantian ethical categories provide no basis for saying that it is more important for individuals to have a right to farm than to have a right to sell encyclopedias, to become doctors, or to operate a business establishment. Indeed, part of the achievement of Enlightenment morality is that it separated moral standing from social role. It should come as no surprise that attempts to apply Enlightenment moral theories to a defense of the family farm become tortured.[5]

Why, then, is family farming singled out for special treatment, and why are masses of non-farmers in industrialized societies willing to spend enormous amounts of public funds to preserve what they perceive to be family farms? The second part of this question has psychological overtones that will not be addressed; the point is to find a moral basis for finding the life of the family farmer special. The most potent thinking on this subject has issued from Kentucky poet and essayist Wendell Berry. The reason that small farms are good is that they cultivate virtue in the character of the farm family. The reasons Berry gives for thinking that farming cultivates virtue do not easily survive condensation and summarization. They have to do with the way that farm families experience the unity and diversity of life. Each member of the family performs diverse roles that are specialized by age and sex. Age and sex are, in turn, precisely the factors that define one's place in the social order of the family. The family unifies these roles into an order that makes each person's duty in assuring farm survival easy to grasp. The diversity of tasks are also reflected in the changing of the seasons and in the breadth of the cultural practices, but these, too, are unified by the farm itself. The farm family is at one with nature, and each person both values and is valued by the role relationships that the production practices of the small farm demands. Similar roles bind all members of the rural community (Berry 1977, 1981, 1987).

What we have in Berry's thought, then, is a revision of the old traditionalism of the feudal system. Moral obligations come forth from roles that unify economic and political status. Duties are grounded in loyalties to particular individuals who are bound to one another in time and place. Virtue is found in living up to the role requirements one inherits by being situated in a community or a family. The particularity of being so situated means that men have roles that differ from women, that adults have roles that differ from children, that farmers have roles that differ from cobblers, blacksmiths, and carpenters, and that each person must fulfill the requirements of his or her roles. Those who do so are virtuous, those who are overcome by jealousy, competitiveness, greed, and other vices fall short in the moral quest. Even the failures know what is expected of them, however, while those of us who live in trading and manufacturing societies have no clear-cut goals to live up to. Berry's most poignant portrayal of the morality of roles is his deeply understated novel *A Place on Earth,* in which a World War II Kentucky farming community perches at the precipice of industrialized agriculture, of sons and daughters who will not take up their parents' roles, and of a community whose inner structure is on the verge of destruction, not by the war, but by the war technologies like DDT that are its aftermath (Berry 1983). Berry's celebration of agrarian virtue is a revision, rather than a revival, of traditionalist morality, because each farmer is lord of the manor. It is a democratization, or at least a leveling, of the feudal class structure, but one which preserves its categories for deriving moral significance.[6]

In fact, Wendell Berry's literary efforts are representative of an attack on the individualism and universalism of Enlightenment morality that has been sounded in other quarters as well. Alisdair MacIntyre's *After Virtue* (1981) and *Habits of the Heart* (Bellah et al. 1985) by five co-authors have also taken up the pen against the way that Enlightenment morality fails to account for the historical and geographical rootedness of moral relationships. Both of these works have been linked to the defense of family farms (Comstock 1987). MacIntyre traces his preferred notion of virtue to the philosophy of Aristotle, and John Lyon offers an Aristotelian reading of Wendell Berry in a 1987 review (Lyon 1987). Communitarianism is the closest relative to neo-traditionalism in the philosophical literature, and it is often taken to be a fundamental and important attack on Enlightenment interpretations of the concept of value (Sandel 1984). Assessing the validity of the communitarian critique of Enlightenment thought is also beyond the present scope. The controversy over the value of the family farm may be a throwback to a notion of value that guided agricultural production in feudal times, and perhaps since the dawn of civilization. A fuller treatment of this issue (and of Wendell Berry's work) can be found in Chapter 5 on rural education, Chapter 10 on the agrarian ideal, and Chapter 11 on constitutional values.

Notes

1. Eric Jones, 1983, *The European Miracle* (Cambridge: Cambridge University Press) argues that the European continent of 1400 to 1800 was uniquely

suited to a kind of moral evolutionism, as societies experimented with new arrangements of individual rights and privileges. Because exit was comparatively easy, groups of people possessing key management and financial skills tended to be quite mobile, moving to the state with the most advantageous social situation. Such states would gain a competitive advantage both in trade and in war. The long-term result was both natural selection and emulation of states that tended to afford key populations relatively greater protection of basic rights and liberties. Jones' essay and the analysis presented here both owe a large debt to Barrington Moore's *Social Origins of Dictatorship and Democracy: Lord and Peasant in the Making of the Modern World* (Boston: Beacon Press, 1996).

2. See, for example, the papers in *Values at Risk,* Douglas McLean, ed. (Totawa, N.J.: Rowman and Allenheld, 1986).

3. See John Rawls, 1971, *A Theory of Justice* (Boston: Harvard University Press), 284-293; Joel Feinberg, 1974, "The Rights of Animals and Unborn Generations," *Philosophy and Environmental Crisis,* W. Blackstone, ed. (Athens, Ga.: University of Georgia Press); *Responsibilities to Future Generations,* E. Partridge, ed. (Buffalo, N.Y.: Prometheus Books, 1981); Derek Parfit, 1983, *Reasons and Persons* (Cambridge: Cambridge University Press); Daniel W. Bromley, 1981, "Entitlements, Missing Markets, and Environmental Uncertainty," *Journal of Environmental Economics and Management* 17: 181-194.

4. See Holmes Rolston II, 1975, "Is There an Ecological Ethic?" *Ethics* 85:93-109; Kenneth Goodpaster, 1978, "On Being Morally Considerable," *The Journal of Philosophy* 75: 308-325; Tom Regan, 1981, "The Nature and Possibility of an Environmental Ethic," *Environmental Ethics*: 19-34; John Rodman, 1983, "Four Forms of Ecological Consciousness Reconsidered," *Ethics and the Environment,* D. Scherer and T. Attig, eds. (Englewood Cliffs, N.J.: Prentice-Hall), 82-92; Edward Johnson, 1984, "Treating the Dirt: Environmental Ethics and Moral Theory," *Earthbound,* T. Regan, ed. (New York: Random House) 336-367.

5. There are several authors who have attempted to apply the categories of modern moral theory to the question of small farms. The best defenses are probably those of James Montmarquet, 1985, "Philosophical Foundations for Agrarianism," *Agriculture and Human Values* 2(2):5-14; and 1987, "Agrarianism, Wealth, and Economics," *Agriculture and Human Values* 4(2 and 3):47-52; Kristin Shrader-Frechette, 1988, "Agriculture, Ethics, and Restrictions on Property Rights," *Journal of Agricultural Ethics* 1:21-40; or Gary Comstock, 1987, *Is There a Moral Obligation to Save the Family Farm?* (Ames, Iowa: Iowa State University Press,), 399-418. Montmarquet and Shrader-Frechette ultimately wind up discussing values that in their view happen to favor family farms but do so for totally contingent reasons. In other circumstances, family farms would have no special claim. Comstock, on the other hand, seems (at the end of his discussion) to favor reasons that I classify as neo-traditionalism in this article. Among those who I think would agree with my assessment of the lack of a

special status for family farms, see Luther Tweeten, "Has the Family Farm Been Treated Unjustly?" in Comstock supra at 212-237 and William A. Galston, 1988, "Should We Save the Family Farm?" *Report from the Institute for Philosophy and Public Policy* 8(3):1-5.

6. John Brewster recognized this component of agrarian morality in his essays, "Technological Advance and the Future of the Family Farm," *Journal of Farm Economics* 40 (1958): 1596-1609, "The Cultural Crisis of Our time," and "The Relevance of the Jeffersonian Dream Today," both in *A Philosopher Among Economists*, P. Madden and D. Brewster, eds. (Philadelphia: J. T. Murphy, 1970) at 7-65 and 175-208 respectively.

Part One

Research

1

Ethics in Agricultural Research

The fact that agriculture is so intimately related to human health and well-being, both at the individual and social level, means that agricultural research frequently aims to improve the human condition. The goal of improvement implies that some sort of normative standard is in mind that allows us to decide which changes in society are to count as better and which as worse. This chapter presents the case that a utilitarian approach to understanding the moral imperatives of agricultural research provides a good starting point for making these decisions and that key research decisions appear to have been made in accordance with utilitarian standards. Utilitarian moral theory has several deficiencies that have been the subject of pointed criticism in philosophy for many years. Sensitivity to these criticisms would enable research planners to avoid pitfalls in applying the utilitarian standard.

The mechanical tomato harvester has come to be recognized as a paradigm example of the controversy that can be generated by agricultural research, and the lawsuit filed against the University of California provides an excellent case study for exploring the ramifications of applying moral standards to agricultural research. This case is complex and might be analyzed with other ends in view. As an example of agricultural research planning, it is important to know that researchers at the University of California made a decision to develop a mechanical harvester for tomatoes intended for processed uses (a decision that also implies development of tomato breeds well-adapted to mechanical harvesting technology). This decision was made on the basis of a judgment that hand labor required to harvest tomatoes would become increasingly scarce in California and that this would, in turn, have two consequences: (1) consumer prices for processed tomatoes would increase, and (2) California producers would face increased risks. The mechanical harvesting technology package was, in fact, developed, and both of these consequences were averted, an accomplishment that was widely hailed at the time (Rasmussen 1968). It has subsequently been suggested that the new technology was instrumental in causing the failure of small- and medium-sized tomato farms, and their replacement by large operations, and in creating unemployment among California agricultural laborers. It is this second charge that was the subject of the California Rural Legal Assistance (CRLA) class action suit claiming damages against the University (Hess 1984). The following analysis will not pass judgment on the accuracy of these suggestions, nor will it discuss the legal merits of the CRLA case. The point here is to examine the type of reasoning that would have lent credence to a positive evaluation of the original research decision, in the first place, as well as the reasoning that would serve as a basis for criticism, in the second.

The mechanical tomato harvester case is relevant to ongoing discussions of research planning because it typifies a pattern of decision and critique that appears to be repeating itself in discussions of agricultural biotechnology. The development of bovine somatotropin was questioned prior to its final approval in 1993. BST was predicted to increase production in dairy cows by as much as 40 percent (actual results are disputed but appear to be lower, though economically significant). While increased productivity can translate into benefits for consumers of milk products and to some dairy producers, it can also affect efficiencies of scale in the dairy industry, forcing current producers to bear adjustment costs and perhaps requiring some to quit the dairy business altogether. A study by Robert Kalter (1985) predicted that many producers would indeed exit the industry, turning the BST case into a politically explosive replay of the tomato harvester controversy.

The tomato harvester and bovine somatotropin are both the end products of extensive agricultural research programs. There is literature on the evaluation of these products and the research programs that produced them, but the existing literature concentrates on consequences that are described as "social," "economic," or "legal." In order to discuss the ethical or moral significance, it will be necessary to make the ethical dimension of these cases explicit and to sketch an outline of the theory of moral responsibility. Each of the two case studies provides abundant illustration of the key points in the theory of responsibility, and both are taken up at recurring intervals in the subsequent chapters of this book.

Ethical Issues in Agricultural Research

Scientific research (and, indeed, almost any human activity) is expected to conform to basic standards of moral responsibility. This simply means that scientists are expected to be judicious in the design and implementation of research projects and to insure that their research does not cause harm. The use of human subjects and the containment of dangerous substances are standard instances in which the moral responsibility of the research scientist would be readily recognized. Generally speaking, research scientists would not be held morally responsible for the social and economic consequences accruing from applications of their research, since it is commonly recognized that research discoveries have both beneficial and harmful uses, as well as consequences and applications that the original research scientists would not be able to anticipate. This traditional standard has been questioned, however, particularly with regard to the research that led to the development of nuclear weapons.

The case with agricultural research can be different in several important respects. First, agricultural research is often done with the advance intention that it be applied in certain ways. Certainly, research on the mechanical tomato harvester was done with the expectation that these machines might someday be used commercially. Second, agricultural research is sometimes initiated to resolve practical problems in agricultural production and may be conducted in close collaboration with producers who understand the research purely in terms of its ability to help them achieve certain specifiable goals. When this is the case, the ethical validity of these goals becomes relevant to the evaluation of research. Finally, agricultural research in the United States has been conducted under the role and scope of provisions of the land-grant university system, which mandates a mission of public utility for agricultural research not universally demanded of scientific research in general (Ruttan 1982). As such, agricultural researchers and their institutions may, in some cases, be held morally responsible for harmful consequences of research beyond those that would traditionally be associated with scientific research. The considerations that allow us to hold agricultural scientists morally responsible for harm are, generally, the same ones that allow us to give them credit when the consequences are beneficial.

Agricultural experiment stations are also expected to conform to standards of public accountability because of their mission orientation and the fact that their research is heavily subsidized by tax dollars. Advocates often justify science in terms of its intrinsic value and the search for truth. Agricultural research, on the other hand, is sometimes expected to perform a public service and is evaluated according to the standards of fairness and impartiality that are more frequently applied to government. Jim Hightower, who subsequently became Texas' agriculture commissioner, subjected agricultural research to a scathing attack on just this basis in the 1970s (Hightower 1978). This aspect of agricultural research might also be formulated as a responsibility to sponsor research that helps meet society's goals (Kaldor 1971).

The belief that agricultural research should, as a matter of ethics, be publicly accountable is the source of several philosophical problems. At the least, it makes agricultural research vulnerable to differing interpretations of "the public good," a notoriously abused concept in American political life (Hadwiger 1982, 75). The key idea seems to be that agricultural research should, in some sense, be "useful" without specifying precisely what usefulness is. Given the fact that public accountability requires agricultural research to be useful, there is some practical value in taking a consequential approach to responsibility. If utility or use value is taken as one of the characteristics that makes an action good, then the usefulness (and hence public accountability) of agricultural research becomes a function of moral responsibility. Such a view of the role of agricultural research has been defended before (Tweeten 1984).

The Utilitarian Theory of Moral Responsibility

If one adopts a consequential approach to the theory of moral responsibility, the rightness or wrongness of an action becomes a problem in evaluating the goodness or badness of outcomes. Utilitarian evaluation of consequences consists in a quantitative comparison of benefits and harms. Drawing heavily on the ideas of Jeremy Bentham, John Stuart Mill offered a utilitarian account of right action in 1861. He wrote:

> The creed which accepts as the foundation of morals "utility" or the "greatest happiness principle" holds that actions are right in proportion as they tend to promote happiness; wrong as they produce the reverse of happiness. By happiness is intended pleasure and the absence of pain; by unhappiness, pain and the privation of pleasure (Mill 1861).

In this passage, Mill interprets benefits in terms of pleasure and harm in terms of pain. This is "hedonistic utilitarianism." More recent theorists have interpreted benefit in terms of whether an individual is satisfied by a particular outcome or state of affairs and harm in terms of dissatisfaction (Griffen 1982). This is "preference utilitarianism," since goodness or badness is determined according to individual preferences. Economists searching for a less subjective approach to the quantification of benefits and harms have adopted the strategy of interpreting both benefit and harm in terms of the economic value of consequences (which can be either positive or negative) under given market conditions. This is "cost-benefit analysis" (Copp 1985). This somewhat simplistic account of utilitarianism is elaborated in Thompson, Matthew and van Ravenswaay, 1992, Chapter 3.

Under a utilitarian interpretation, an agent will be held morally responsible for both benefits and harms, will be praised to the extent that benefits outweigh harms, and will be blamed or condemned to the extent that harms outweigh benefits. In addition, agents must act in a way that brings about the greatest balance of benefit and harm, so that an agent who foregoes an action that might have produced more net benefit than the one actually performed may be criticized for failing to maximize. We might say that the utilitarian theory of responsibility evaluates an act according to two central criteria:

> (1) Productivity. Does the act produce benefits? Given that harms are interpreted as negative production, or consequences that negate the value of benefits produced, the net productivity P of the action is the result of benefit B minus harm H so that $P = (B - H)$. An agent is responsible for the benefits and harms the action produces.
>
> (2) Efficiency. Does the act maximize the benefits produced for all people for each effort expended? Acts that fail to produce maximal benefits are irresponsible.

Economic analyses of the mechanical tomato harvester have applied technical interpretations of these two criteria to evaluate the agricultural research program that produced it. Two separate studies have shown that benefits outweigh harms, when measured in economic terms (Schmitz and Seckler 1970; Brandt and French 1983). These evacuations, interpreted within the framework of the utilitarian theory of moral responsibility, would place agricultural research in a very favorable light. Utilitarian theory, however, has several widely acknowledged deficiencies, three of which are discussed below.

Deficiencies in Utilitarian Theory: Equity

The economic studies cited above both note that, although the tomato harvester increased production efficiencies in the tomato industry, thus lowering the cost of tomatoes to consumers and producing a very general and considerable benefit when this result was aggregated, the harms associated with its commercial introduction were not widely distributed and were experienced by a fairly narrow range of agricultural laborers and small producers. Schmitz and Seckler (1970) see the problem as one arising from the powerlessness of these groups, which prevented them from seeking appropriate compensation from those who benefited. Brandt and French (1983) suggest that improved wage rates and working conditions in non-harvest sectors of the tomato industry may have produced partial compensation to workers but concede that "consumers have been the primary long-run benefactors [sic] of this technological development" (271).

The utilitarian theory of moral responsibility does not include any criteria for distributing benefits and harms. If the total aggregated benefits to consumers of saving a few cents on tomatoes is greater than the total loss of income suffered by displaced workers, the utilitarian principle will support the development of the harvester, despite the fact that the degree of harm experienced by each displaced worker far exceeds the amount of benefit experienced by any single consumer. The utilitarian theory has no mechanism to guard against an inequitable distribution of benefits and harms. Utilitarian philosophers have attempted to resolve this problem through more emphasis on rule following (as opposed to maximization) and on the theory of fairness (Rescher 1966; Rawls 1971). Economists have attempted solutions that range from prediction of distributive impacts to Pareto improvement criteria (Madden 1986). Some see distributional problems as a reason to reject utilitarianism entirely (Dworkin 1977; Machan 1984). Each of these solutions involves considerations that go beyond the scope of this paper.

Deficiencies in Utilitarian Theory: Sustainability

A second problem with the utilitarian principle of maximizing benefits is that its scope is not specified precisely. It is impossible to anticipate all the benefits and harms that might follow from an action, particularly when the action may have long-range impact upon methods of agricultural production. Interpreted strictly, the utilitarian maxim would require one to maximize benefits and minimize harms for all persons at all times, but the impossibility of doing this means that, in practice, one must adopt a (perhaps) arbitrary limitation of scope. An action that will contribute to severe harms ten, twenty, or even a hundred years in the future may be justified by the utilitarian approach to responsibility when the benefits and harms of the immediate present are the basis of decision (MacIntyre 1977).

This problem has been raised in connection to agriculture with respect to the question of sustainability. The mechanical tomato harvester is not typically cited as a prime offender, but critics of the agriculture research establishment have blasted mechanization as an inappropriate attempt to "apply the principles of industry to agriculture" (Schumacher 1973). More moderate critics note that attempts to maximize short-term productivity and efficiency create economic disincentives for conservation practices (Batie 1984). Madden (1984) notes that, although there is a gulf between the rhetoric of sustainable agriculture and conventional agricultural research, there is considerable overlap both in actual farming practices and in beliefs about the importance of long-range consequences.

As such, it may be that this deficiency of utilitarian approaches to responsibility can be moderated by diligent attention to long-range impact upon benefits and harms of agricultural research.

Deficiencies in Utilitarian Theory: Autonomy

In attempting to analyze why the tomato harvester became such a controversial topic, Thompson and Scheuring (1984) write that,

> The results of technology seem to many critics to result in a less human life, i.e., a physical and social environment in which human beings are increasingly cut off from the natural world of which they are biologically a part (145-146).

Their point is that mechanization is seen by some as part of an essential conflict between those who evaluate technology in terms of net social benefits and those who see it in conflict with basic human values. Indeed, much of Jim Hightower's criticism of the consequences of agricultural research is not directed against distributional or even ecological impacts of new technology. More frequently, he is attacking what he sees as technology's effect on the farmer's way of life. Technological advances create competitive advantages for those who use them; thus, one has no choice but to use them and to adjust one's life to the dictates of the technological imperative. In this sense, the restriction of freedom or the loss of autonomy is seen, by some, to be one consequence of agricultural research (Hightower 1975).

Clearly, autonomy, or the capacity of individuals to choose for themselves, does not figure in the utilitarian goals of productivity except to the extent that it contributes to pleasure, satisfaction, or the pursuit of economic well-being. While it is plausible to assert that autonomy is vital to a pleasurable life and to the satisfaction of preferences, it is difficult to see how economic interpretations of cost and benefit can accommodate this value, and this is one reason that some authors reject entirely economic interpretations of utilitarianism (Sagoff 1986). Still others cite this as reason for giving up on all versions of utilitarianism, opting instead for a theory that defines moral responsibility in terms of the absolute and uncompromising respect of other people's rights to life, liberty, and pursuit of happiness (Machan 1984). Again, the resolution of such difficulties becomes quite detailed.

Moral Responsibility in Agricultural Research

To the extent that agricultural research is planned and directed toward the goals of increased productivity and efficiency, the utilitarian approach to the problem of moral responsibility is already implicit in agricultural research. Of course, there are some areas of basic research in which utility is not currently a primary consideration. Social utility is not normally a major criterion for the evaluation of research in the pure sciences, or the humanities, and it seems appropriate, given the goals of the university, that it should not be exclusive criterion for agricultural research either. To the extent that it is accepted as a goal for agricultural research, however, the planning and conduct of agricultural research is already committed to a vision of moral responsibility. There is no need to call for ethics in agricultural research. An implicit ethic has been there all along.

The utilitarian approach to moral responsibility is not without some deficiencies, however. Areas of deficiency in utilitarianism are closely correlated with criticisms of agricultural research over the last two decades. The mechanical tomato harvester is particularly associated with problems in the inequitable distribution of harms and benefits and with its impact on the autonomy of individual decision makers. The mere fact that such criticisms have been raised does not, of course, mean that they are valid, but the fact that they are all representative of well-known deficiencies in utilitarian theory should give agricultural researchers reason to reevaluate the implicit ethic that guides agricultural research. It might be possible to formulate a "checklist" of ethical concerns that are not well represented under utilitarian criteria.

If we were to apply this model to the evaluation of biotechnology research in general and to bovine growth hormone in particular, we would expect to find emerging concerns in the three areas of equity, sustainability, and autonomy. In fact, all three concerns had emerged well before the political controversy over BST began to boil. Robert J. Kalter (1985) writes:

> Although forecasts of this nature are often dangerous, the economic trends that will result from the modern agricultural biotechnology can already be discerned. At the farm level there will be clear winners and losers. In the middle will be a large proportion of farm operators for whom this new technology promises major challenges if they are to continue in farming (129).

Kalter's study focuses on the equity problems associated with bovine somatotropin. Although BST was not frequently linked to environmental issues, many of the first criticisms of biotechnology raised concern regarding sustainability and long-term impacts on the environment (Alexander 1985; Rifkin 1985; Doyle 1985). Finally, Jack Kloppenburg (1984) made the following criticism of recent biotechnology research:

> Like hybrid corn before it, biotechnology should stimulate an increase, rather than a decrease, in the intensity of chemical usage in agriculture. ... In doing so it will also follow the pattern set by hybrid corn ... : the progressive erosion of the autonomy of the farmer and his increasing dependence on factor markets. As plants and animals become increasingly "programmed," they will require sophisticated monitoring and management packages if their productive potential is to be realized.

Kloppenburg is expressing a concern that a decreased freedom of choice and control for the individual will be a consequence of biotechnology research. Even a cursory review of the literature, therefore, reveals that each of three deficiencies in the utilitarian model cited above have already given rise to criticisms of agricultural research on biotechnology.

Although this discussion is intended solely as introduction to some of the cautionary principles that ought to constrain application of a utilitarian ethics of responsibility in agricultural research, it is worth mentioning in passing that the issue of responsibility for the impact of agricultural research may be determined by factors that have not been discussed above. There should also be some attention to the structural elements of moral responsibility (e.g., agency, causality, and intention) as well. Indeed, some of the most recent work on the ethical significance of the tomato harvester has focused primarily on the issue of causality: Did the University of California's research cause the changes in the tomato industry that are the basis of the CRLA lawsuit? A recent study indicates that most of these changes would have happened anyway (Martin and Olmstead 1985); if the research did not cause the changes and their harmful effects, then the researchers and the research institution cannot be held responsible. This argument in itself, is not enough to prove moral innocence, however. The murderer who puts a bullet through the skull of a terminal cancer patient cannot plead to not causing the death because the victim would have died anyway. It is important to have a clear understanding of how research institutions act and what events their actions can cause if we are to understand the limits of moral responsibility in agricultural research and to know when these research institutions cannot be held responsible for structural reasons.

Conclusions

The utilitarian model of moral responsibility accounts for key elements of the ethical implications of agricultural research. It illuminates when agricultural researchers and institutions can take credit or blame for important benefits and harms of their research. The emphasis on productivity and efficiency in the utilitarian model of moral responsibility provides a way of accommodating the belief that agricultural research has a duty of public accountability, as well as more standard (and less onerous) requirements of moral responsibility. Furthermore, the utilitarian model seems to be implicit in many agricultural research decisions.

The utilitarian model is not without several well-documented philosophical deficiencies with respect to moral responsibility. In particular, the model does not generally provide an adequate basis for evaluating equity, sustainability, or autonomy when these concerns are deemed relevant to the situation at hand. There is a correlation between these areas of deficiency and criticisms that have been raised against agricultural research. Research planning in the future might profit from a more comprehensive consideration of the deficiencies in the utilitarian approach.

2

Elements of Ethical Choices for Agricultural Research

The utilitarian approach to understanding the moral imperatives of agricultural research offers a good starting point for making decisions about which agricultural technologies to develop and support, but this endorsement of utilitarianism must be qualified. Utilitarian moral theory has deficiencies that are well known. Sensitivity to them would enable research planners to avoid pitfalls in applying the utilitarian standard. In this chapter, the recognized deficiencies of utilitarian theory will be further explored. The final section will provide more specific suggestions about how to be sensitive to deficiencies in utilitarian theory and will discuss some of the obstacles to improved ethical decision making for agricultural research.

Utilitarian Research Planning

The first chapter provides a cursory overview of utilitarian moral thinking. For present purposes, any approach to decision making which follows a simple four-stage process will be understood as utilitarian.

1. Define the decision as a set of discrete or mutually exclusive options.
2. Predict and represent the likely consequences associated with each option. This assessment can be quite complex and will, in sophisticated approaches, assume that multiple outcomes can be expected with respective degrees of probability. However complex and probabilistically anticipated, each option will have an outcome set of predicted events associated with it.
3. Assign value to each outcome set. The assignment of values to expected outcomes can be relatively easy when outcomes represent fi-

nancial loses and gains, and can be quite difficult when outcomes include impacts upon health, environmental quality, or access to other goods not traded on markets. Value assignments need not be cardinal, meaning that outcomes may be ranked in order of preference, only.

4. Apply a decision rule which operates on the values assigned or expected for each option. The decision rule selects which option is the right one.

This account of decision procedure is admittedly abstract. Johnson (1982, 1984b) has offered a more detailed and systematic account that illustrates how each of these stages can be reiterated, and how positive and normative information interact in the interaction. Johnson also stresses how decision continues through stages of execution and responsibility. The point here is to provide a very general account of decision making that might be made specific and applied to agricultural research in several different ways.

For example, the decision to release a genetically engineered variety of cotton can be fitted to these four stages. At a minimum, one has the options of releasing the cotton or not releasing it. There may be more subtle ways to define the options available, and they might include forgetting about the cotton entirely and taking up a new career in ballroom dancing. Hence, the first stage is to define the options one will entertain. The second stage is to predict the outcome of each option. Here, it will be important to know whether the cotton will be adopted by producers, and at what rate. It is important to know what, if any, environmental impact adoption of the cotton will have. In the third stage, impacts identified in the second stage are evaluated. In classical utilitarian theory, they would be evaluated in terms of their capacity to produce pleasure or pain for sentient creatures. In a formal benefit-cost analysis, the value of consequences must be monetized. Although there are many ways to assign values to consequences, the standard practice in agricultural research planning has been to ask economists to estimate money values associated with profitability for cotton producers and reduced costs for cotton consumers. Only recently have health or environmental impacts been included in the calculation. The final stage involves the application of a decision rule. Classical utilitarian theory specified an optimizing rule called the utilitarian maxim. The rule states that the decision maker should select the option expected to produce the greatest value for the largest number of people. If applied to a two-option cotton variety decision, the utilitarian maxim says to release the variety if doing so produces greater benefit than not doing so.

There are different ways to flesh out each stage. One could, for example, use a decision rule that selects the option with the greatest personal value for the decision maker, regardless of consequences for others and still remain consistent with the four stage process described above. Such a rule would not win praise for its sensitivity to ethical concerns. Stated in such general form, these four stages provide an extremely broad interpretation of the utilitarian framework for decision making. The stages are broad enough to incorporate most approaches to decision making that would be consistent with decision theory (Giere, 1991). The four stage framework for understanding the ethics of decision making derives from a tradition of utilitarian philosophy that would typically interpret each stage in a specific way. A utilitarian would insist:

- that all meaningful options be identified in the first stage;
- that consequences be predicted for all affected parties in the second stage;
- that the means of evaluating outcomes in the third stage is impartial and does not produce higher valuations for impacts to parties favored by the decision maker in virtue of race, sex, religious affiliation, personal relationship, or sexual preference; and
- that the decision rule be an optimizing one which accounts for impacts to all affected parties.

It is reasonable to assume that decision makers in agricultural science have attempted to interpret the four stage framework in a manner consistent with these four additional utilitarian qualifiers.

Historically, scientific research has not been subjected to this four stage test for ethical defensibility, but agricultural research is the exception. Agricultural research has differed from general scientific research in that there has always been an assumption that agricultural researchers should be accountable to the public. The utilitarian model of decision making has helped researchers justify decisions leading to increased yields of agricultural commodities because these increased yields were widely thought to produce benefits in the form of higher farm income and lower food prices for consumers. Research administrators may not have conducted detailed formal applications of the four stage process in reaching these choices, but there can be little doubt that a simple and straightforward application of the procedure would have indicated many of the choices that they actually made (Thompson and Stout, 1991). In this respect, the suggestion of a "Fourth Criterion," (Lacy and Busch, 1991) merely makes explicit what has always been assumed for agriculture.

Chapter 1 notes that challenges to agricultural research have paralleled several acknowledged deficiencies of utilitarian theory. The first of these is equity. To the extent that one simply maximizes net benefits, one may select options which produce a lopsided distribution of benefits and costs. In agriculture, this objection has taken the form of statements to the effect that research favors large producers over smaller ones. The second deficiency is sustainability. The utilitarian decision maker may simply fail to foresee long-term costs associated with research. If the new cotton variety will eventually cause cotton pests to acquire resistance to important and ecologically sound pest control agents, for example, costs might outweigh benefits over the long run.

The third deficiency is autonomy. The utilitarian decision maker may make choices which foreclose another individual's opportunity to choose for herself. The utilitarian decision maker may see the release of a new variety of cotton as a path to the greater good, but, in the irony of the market, the availability of the new variety may restrict, rather than expand, the choices available to producers. This restriction of choice may be a necessary "cost," to achieve the greater good. A cost that imposes a restriction in freedom for others may be more than a simple cost, however. It may be an instance of using another person merely as a means to achieve the greater social good.

To some extent, each of these deficiencies can be moderated or avoided through a sophisticated application of the four stage process. It is certainly possible to include distributive impacts among the consequences that are predicted, and to assign values that reflect positively or negatively on distributions that are skewed in egalitarian or elitist directions. Furthermore, though practical necessity demands some limitation of scope, expansion of scope to the genuinely long run promises to make utilitarian decision making more sensitive to sustainability. Finally, it is possible to adopt decision rules which refuse choices that interfere with autonomy. This last deficiency, however, will prove among the most difficult to moderate. It provides a good opportunity for taking up a more critical look at the utilitarian framework.

Objections to a Utilitarian Framework

Even without explicitly utilitarian interpretations of each element, the four stages in the framework discussed above have at least two highly significant results for ethics. First, rules (1) and (2) do not seem to involve ethics in any obvious way at all. The decision problem is divided into an information gathering stage and an evaluation stage. The information gathering stage appears to call for empirical research, while the evaluation stage calls for a combination of economic, sociological and ethical analysis. Second, within the evaluation stage, it is future states of affairs, what Nozick (1974) called end states, that have value. Ethics enters in the determination, measurement and assignment of that value and in the selection of a decision rule which must take those value assignments as its input variables. Sensitivity to deficiencies in the utilitarian framework requires special attention to these two results.

As noted, the utilitarian framework implies an impractically idealistic requirement to consider every consequence of an action. In real life, one must limit the time frame in which consequences are calculated, and an inappropriate limitation of that time frame can threaten sustainability. Similarly, the utilitarian interpretation of the first stage implies that all options will be considered. This, too, is an extremely idealistic requirement, since at any given decision point, a decision maker has an infinite number of frivolous options, as well as some viable options that may easily be overlooked. In point of fact, the decision maker must use judgment in both stage 1 and stage 2 in order to limit both the number of options to be considered and the time frame for which consequences will be predicted. Such judgments require the application of norms and values. The decision maker brings values to the first stages of choice that determine appropriate parameters for these judgments (Thompson 1995). The assumption that stages 1 and 2 do not involve ethics is therefore mistaken. This is a point that has been well noted by Glenn Johnson (1982, 1984a).

Furthermore, the assumption that stages 1 and 2 do not involve ethics is also very likely to produce problems with respect to autonomy. The utilitarian framework threatens autonomy in several ways. One is that new technologies may foreclose opportunity, but a potentially more serious challenge to autonomy can arise from applications of the framework itself. Producers, consumers and others affected by new technology are objects of study for someone who attempts to apply the utilitarian framework. The impacts upon them are assessed and evaluated. There is nothing in the framework, however, which assures or offers them any role in the actual decision-making process. A sophisticated use of the framework will seek input from affected parties in order to assess the value of predicted consequences, but while this input may improve the analysis and help make the choice more palatable, it does little to restore autonomy. If affected parties have not had an opportunity to help select the options to be considered and the consequences to be predicted, decision makers are functioning as gatekeepers at these stages.

This challenge to autonomy is heightened further by the fact that end states are taken to have value. This practice effectively makes affected parties into vessels that contain quantities of satisfaction or dissatisfaction, cost or benefit. Other human beings are understood as passive recipients who experience the consequences of choice. The choice of an option is the cause, and the consequences are effects. As affected parties, rather than affecting agents, other human beings are cast in a role which denies their agency, their ability to choose and act. A utilitarian decision analysis is, thus, liable to become highly insensitive to the characteristic of human beings that may be most highly prized and most closely associated with an individual's personal identity. Respect for autonomy, understood as the capacity to choose and act of one's own free choice, simply cannot be assigned value in an end state, precisely because treating people as if they are only affected parties, not autonomous agents, is already a violation of autonomy.

These points can be brought into relief by considering a thought experiment originally offered by Bernard Williams. This thought experiment, known as "Jim and the Indians," is useful for illustrating deficiencies in utilitarian thinking.

> Jim finds himself in the central square of a small South American town. Tied up against the wall are a row of twenty Indians, most terrified, a few defiant, in front of them several armed men in uniform. A heavy man in a sweat-stained khaki shirt turns out to be the captain in charge and, after a good deal of questioning of Jim, which establishes that he got there by accident while on a botanical expedition, explains that the Indians are a random group of the inhabitants who, after recent acts of protest against the government, are just about to be killed to remind other possible protesters of the advantages of not protesting. However, since Jim is an honored visitor from another land, the captain is happy to offer him a guest's privilege of killing one of the Indians himself. If Jim accepts, then as a special mark of the occasion, the other Indians will be let off. Of course, if Jim refuses, then there is no special occasion, and Pedro here will do what he was about to do when Jim arrived, and kill them all. Jim, with some desperate recollection of schoolboy fiction, wonders whether if he got hold of a gun, he could hold the captain, Pedro and the rest of the soldiers to threat, but it is quite clear from the setup that nothing of that kind is going to work; any attempt at that sort of thing will mean that all the Indians will be killed, and himself. The men against the wall, and the other villagers, understand the situation, and are obviously begging him to accept. What should he do? (Smart and Williams, 1973, pp. 98-99)

In its original context, this thought experiment is introduced to raise questions about whether utilitarian ethical theory can deal adequately with questions of moral character, especially when the virtue of integrity is at stake. Contrary to Williams' implied suggestion, I do not think that contrived thought experiments which pose dilemmas of this sort reveal very much about any individual's core beliefs or character, nor do I think that abstract philosophical considerations can tell us very much about how a person would or should act in a real crisis situation.

Jim's choice is so overwrought with pathos and drama that it is an exceedingly poor example of ethical decision making. Jim and the Indians exaggerates several common elements in ethical decision making in order to promote drama. While this exaggeration renders the experiment highly artificial, it also makes these elements more obvious than they are in cases where choices over agricultural technology must be made. Jim and the Indians is, for this reason, a useful review of common problems encountered in the four stage model of ethical decision making.

First, if we apply a typical interpretation of the four stage model, there can be little doubt that Jim must kill an Indian. The example limits Jim's options and so thoroughly specifies the consequences of either refusing this act, or of attempting a rescue, that alternative ways of interpreting the third and fourth stages seem highly unlikely to recommend any other course of action. A sophisticated economic analysis of Jim's options may help him minimize the harm done by indicating which life is worth the least, or how the preferences of all parties (most especially the Indians) may be most efficiently satisfied. He may, for example, choose an old or sick victim with no dependents, or he may ask for volunteers. The considerations that might be introduced to make more optimal choices will not change the judgment that one death is better than twenty. It is the idea that choices should be made on the basis of their predicted outcome that determines the choice here, and the details of decision theory and economic analysis are window dressing. If Jim should base his decision on the predicted or expected state of affairs that will follow, the ethics of the choice are, in one sense, easy.

Williams notes that many people will at least feel qualms about the ease of arriving at the decision to kill an Indian, and in a series of unscientific samples of individuals to whom I have suggested similar choices over the years, 30 to 50 percent refuse to kill an Indian outright. A significant minority report a preference for suicide, despite the lack of any obvious benefit such a course of action could produce for the Indians. I repeat that I do not take these refusals to represent meaningful evidence about the ethical values of individuals in question. The unease with which many people regard what, on utilitarian grounds, is an easy choice does point toward the conclusion that something other than the consequences of a decision is relevant to an ethical evaluation of Jim's choice. This is the point made obvious by exaggerating and manipulating the circumstances of choice. While we may be convinced that Jim really should kill an Indian, the ease with which a true utilitarian reaches this recommendation is disturbing.

There are many points in Jim and the Indians that might be sources for unease. A review of the philosophical literature on the case would take us far afield and would raise issues unrelated to agricultural research. Notice that Williams has set up the case so that the options and consequences of options have been fully specified. Furthermore, Jim is in a position where he must regard the Indians as vessels of value. Jim must ultimately use the life he takes as a means for saving nineteen others. The choice can be easily rationalized as a trade-off of costs and benefits, and Jim may weigh the costs and benefits of one Indian death against another. The Indians are affected, rather than affecting, parties. For that matter, Jim himself is described as an affected party. In constraining the problem definition to one in which both Jim and the Indians have been denied any opportunity to act as free agents, Williams has produced a thought experiment that forces us to conceptualize an issue in thoroughly consequentialist, utilitarian terms. The denial of autonomy that must accompany this contemplation is almost certainly a component of the unease with which one entertains the available options.

Toward More-Sensitive Decision Making

One way to think of ethics in agricultural research planning is to think that decision makers must simply do a more thorough job of anticipating consequences, and of assigning value to them. This approach suggests a need to predict and assess social impact in the second and third stages of decision analysis. Social impact assessment would certainly help address matters of distributional equity and would presumably take up community health and quality of life, as well. The ability to address such concerns would improve decision making. Nevertheless, if the decision maker, whether an individual or a committee, simply factors social impact assessments into the set of expected consequences associated with a research or technology option, there is little reason to think that these decisions will be more sensitive to autonomy than current ones. Even when social impacts are assessed, the human beings who will experience these impacts are being treated as affected parties, rather than active agents, and they are not part of the decision regarding which options and consequences to assess.

To some extent, calls for social assessment are a response to the criticisms leveled against agriculture during the past two decades and an attempt to improve decision making for the potentially controversial development of biotechnology in agriculture. However, there is reason to doubt that more inclusive assessment of ethically significant consequences will be sufficient to assure orderly acceptance of technology. The cases with which we have already had experience provide ample illustration of why this is the case. No case raises the issues of autonomy better than the case of recombinant bovine somatotropin (BST).

The literature on BST is large and continually expanding. Ethical criticism of this technology has focused on three kinds of impact:

- economic studies have predicted distributional impacts upon the structure of the dairy industry;
- animal welfare activists have raised concern about the impact of BST on dairy cows;
- consumer interests have raised questions about food safety and have demanded labeling of BST milk.

These are precisely the sorts of impact that would be addressed through more sophisticated attention to some of the known deficiencies in utilitarian decision making. Impact upon farm structure is a distributional question, dealing with equity. Impact upon animals can be easily accommodated in the framework by expanding the scope of consequences to be predicted and evaluated. Conceived of as a simple risk issue, food safety is among the most thoroughly researched and anticipated of any food technology's expected impacts.

Scientists did an admirable job of attempting to predict and evaluate these consequences. The result, however, was far from an orderly decision process for BST. Instead, interest groups representing farm groups, animal welfare activists and consumer concerns have succeeded in raising broad public fears over BST and biotechnology in general. If any technology has been well studied and anticipated, both through clinical trials and through basic science knowledge of biochemistry, it is BST. Why did BST become a difficult political issue?

There is no shortage of potential answers for this question and a lengthy analysis follows in Chapter 9. Some have noted the symbolic importance of milk as a factor influencing consumer protection of risk. Others have noted the worldwide surplus of milk. Still others have noted the lack of any substantial health or quality benefit for consumers. Each of these factors may be at work, and each would suggest still more impacts to be predicted and assessed in a decision analysis. There is, however, an alternative explanation which gets more directly to the heart of ethics in technological decision making.

In the development of BST, dairy producers, dairy cows, the environment, and milk consumers have consistently been treated as affected parties. The impacts upon them have been assessed in an open and objective manner, but they have been taken as receptacles of value, as the bearers of impact, never as agents with active interests in the course of affairs. It should be obvious that dairy producers and food consumers are potentially agents, and animals and the environment have plenty of human spokespeople willing to serve as agents for them. While key decision makers in both public and private sector organizations may have honestly and objectively attempted to factor impact upon these interests in their decision making, by using the utilitarian (or consequence valuation) framework for decision, they adopted a particular philosophy about democratic decision making which implicitly cast these groups into the role of patient, rather than agent.

Alternatively, one might have started from the fact that BST was developed intentionally, and on purpose by scientists and managers in large organizations. Since the decision was undertaken intentionally, impacts upon affected parties are not experienced as the result of natural causes or the invisible hand of economic change. As such, they are not experienced as impacts which must simply be accepted and for which no one can be held responsible. Affected parties might well be justified in reacting critically to the development of this technology if either of two questions is answered affirmatively.

1. Do those who research and develop technologies like BST have unequal advantages over those who bear the unwanted consequences?
2. Do researchers who developed BST have a charge to protect the interests of those who experience unwanted consequences, regardless of overall value of consequences for society as a whole? (Thompson, 1992, p 37)

Chapter 9 offers a fuller treatment of these questions, but the analysis there can be anticipated in the present context of examining deficiencies of utilitarianism.

A non-utilitarian response to the problems raised by BST would have been to invite all affected parties into the earliest stages of decision making, offering them genuine partnership in the decision making process. Followed seriously, such an approach would respect the autonomy of all parties, restoring them to a position of collaboration, choice, freedom and agency within the decision-making process. One can also anticipate that the bargaining and consensus seeking that would be necessary to reach agreement would be time-consuming, at least, and would not guarantee that objectives such as efficiency, improved quality of life, sustainability, or even equity would be met. As such, a *fully participatory* decision process sacrifices all other ethical values to the alter of autonomy. It is unlikely that this is the solution to more-sensitive research decision making.

In fact, many of the parties who would be invited to the table for participatory decision making would prefer to spend their time at other things. What they do want is to feel secure in the knowledge that their interests are being fairly considered. In short, they must be in a relation of trust with the decision makers who will consider the options, consequences, values and decision rules to be applied in making a decision. To the extent that one trusts both the decision maker and the process, one may be quite willing to leave a technology decision to a person or group using the four stage decision process. Once that trust is eroded in even the slightest degree, however, there is nothing in the four stage process that points decision makers back to a procedure that will reestablish a trustworthy, participatory decision process. This point has been well noted by authors reviewing the role of risk communication in public decision making, (Sandman, 1985; Slovic, 1991; Covello, Sandman and Slovic, 1991) and many of the rules and procedures noted in that literature should be reviewed by agricultural research planners.

Conclusion

Agricultural research planning has a strong ethical pedigree in its century-long application of utilitarian criteria. However, the recent history of agricultural technologies such as the tomato harvester and BST provides ample illustration of how theoretical deficiencies in classical utilitarian thinking can manifest themselves in practice. Research and technology planning can be made more sophisticated by noting several flash-points where the utilitarian model can fail. These include equity and sustainability, along with a need to include a full accounting of costs in the form of health, safety, animal, and environmental impact. There is still a need for careful thinking on how to quantify and compare these impacts. However, increased sophistication in utilitarian decision analysis must not lead analysts to become insensitive to the larger issues of autonomy, participation, and democracy.

Decision makers must do more than remain receptive and responsive to criticism. They must actively seek participation from all sectors of society, even when doing so results in short-term inefficiencies. The long-run efficiency of science and technology planning depends upon relationships of trust that are built slowly and can be expended suddenly and with disastrous results. Too much emphasis upon technical consequence assessment diverts energy from consensus seeking and participatory planning. The need for trust is a common element in public decision making, and trust, ironically, cannot be assured merely by making the right decisions. Sometimes it can be more important to make the wrong decision in the right way.

3

Technological Values in the Applied Science Laboratory

The preceding chapters introduce the question of ethics as related to research choice and provide a moral vocabulary for evaluating research decisions in light of utility and rights. A central claim of both chapters is that while utilitarian moral philosophy provides a good starting point for prospective evaluation of agricultural technology, it is also the case that criticisms of agricultural research correspond to known weaknesses in utilitarian approaches to ethics. This chapter frames the philosophical problem that is implicit in the previous two. It is a problem that has received little attention from anyone and almost none from professional philosophers. Applied science wears its allegiance to the goodness of curing disease, building bridges or producing food on its sleeve, yet studies that have raised questions about the social impact of medical, engineering, or agricultural technologies have, for the most part, been reluctant to press those questions at the point of choosing to conduct research that was intended to create precisely the technologies and techniques that are questioned. If we expect to impede or direct technological change, it is immanently reasonable to think that the most effective point of intervention will be at the stage of research choice.

Such an intervention need not be anti-technology. Indeed, attempts to influence research choice are likely to take shape by way of directing researchers to produce technology, though perhaps technology of a different sort than is produced now. There have been several moments in the past two decades when detectable amounts of political enthusiasm for research on "alternative" technological goals have emerged. There have been calls for research on alternative fuels and on recycling, for example. It is not obvious, however, how such calls for applied research are to be justified philosophically, nor is it obvious that scientists have a responsibility to respond. Any call for changing the research agenda (or for resisting change, for that matter) demands an argument. But what sort of argument, and to whom should it be directed?

There are several philosophical questions to be posed with respect to the problem of research choice, but this chapter takes up one that reaches into the institutional structure of agricultural science. Although many agricultural scientists work in private or commercial development groups dedicated to the production of marketable technologies, many work in non-profit laboratories and universities that are organized around disciplinary departments such as physiology, entomology, or engineering. Virtually all agricultural scientists are trained in such a structure. Is it obvious that disciplinary organization in agriculture serves the explicit social goals that agricultural science is expected to achieve? Does the disciplinary structure of agricultural science itself impose directionality on technical change and, hence, upon fundamental political issues? If so, what, if anything, should be done about it? These are the questions that this chapter attempts to place into a philosophical context. The goal is to give the question a philosophical shape and to prompt a wider reflection upon them. Answers will not be forthcoming.

The Political Economy of Research Choice

In 1983, Lawrence Busch and William Lacy published the results of their sociological study of research choice in agricultural science. Busch and Lacy were attempting to show how granting agencies, tenure and promotion, peer review and influence exerted through state legislators, farm organizations, and agribusiness shaped the agenda for research in agriculture. Their aim was to develop a "political economy" approach to the sociology of science that would contrast with Mertonian models that stressed organization of scientific disciplines around shared theoretical assumptions, research methods and discovery goals (Busch and Lacy 1983). While there has been a significant appreciation of the philosophical implications of the Busch and Lacy study among those interested in agriculture, it has not been widely related to the philosophy of technology. Some philosophers of technology, following Heidegger, have stressed the ontological or historical priority of technology over science; others have analyzed the way in which existing technology imposes an institutional framework upon the activity of scientists. Both of these themes share Busch and Lacy's rejection of science as a pattern of inquiry guided autonomously by scientists in pursuit of rigor, parsimony and truth.

Nevertheless, by focusing on an applied science devoted to the solution of practical problems, Busch and Lacy's work demands that we examine a link between philosophy of science and philosophy of technology that is both more obvious and more subtle than these.

In being configured to solve problems in manufacturing, in agriculture, or in medicine, applied sciences are, by definition, intended to produce technologies such as materials, machines, plant varieties, drugs, and surgical techniques. There is no possibility of even a pretense that the organization and conduct of applied research is *purely* an attempt to describe or explain the world accurately. Applied sciences are geared toward the development of technology from the outset, at least in so far as technology is understood as knowing "how to do it," whatever "it," is. The political economy approach to the sociology of science suggests that the choice of which technologies to develop is determined by "material" forces such as the granting agencies and interest groups already mentioned. As such, it opens a direct route to philosophical questions about distributive justice, social efficiency, human rights and violations of liberty as they relate to scientific research. If research choice is determined by "the forces of production," it is important to raise ethical questions about how applied science serves to concentrate and distribute social power. The political economy approach shows why the research agenda is simultaneously a technological choice and a political choice, at least for the applied sciences.

The political economy approach also suggests that change in research directions can be precipitated simply by a realignment of the rewards system for applied scientists. If applied science follows money (to oversimplify), the direction of research should be unilaterally correlated with availability of funds. By contrast, I will argue that technological values operating within applied science laboratories constrain the influence of the rewards system upon problem choice. While I think that what I will say is true for applied science in general, the main impetus for my reflections is the current state of agricultural research. Over two decades, agricultural research has been subjected to repeated criticism for emphasizing productivity and yield-enhancing technology at the expense of environmental quality, small farm and rural community development, human health and safety, and just, sustainable development in the Third World. While there was some initial resistance to these criticisms, they have now been very substantially accepted by scientists and administrators within most major research organizations and funding agencies. During the past decade the alignment of political interest groups has also shifted to substantially favor alternative streams of agricultural research. Streams that are applied not to the enhancement of productivity, but to the goals advocated by the critics of the 1970s. Despite these shifts, the evidence is that even those research projects nominally dedicated to low input or sustainable alternatives continue to be dominated by the productivity paradigm.

Productivity as a Value in Agricultural Research

The agricultural scientist's emphasis upon productivity is ethically defensible on prima facie grounds because increases in food availability obviously serve the comprehensive ethical norms of social benefit. Productivity goals help shape a researcher's choice of topics in much the same way that the goal of ending disease shapes the medical researcher's choice of topics. It is, in other words, a norm that provides a certain structure to the agricultural researcher's activity and provides a rolling horizon for measuring the success or failure of any given research stream. It is quite common for language expressing an experiment's or research finding's contribution to agricultural productivity to be included in grant proposals, in tenure and promotion reviews, in awards recognizing the achievements of a particular researcher, as well as in technical publications themselves. My book *The Spirit of the Soil: Agriculture and Environmental Ethics* (1995b) presents an extended discussion of how this norm can be built up into a comprehensive (if indefensible) philosophy of agriculture which I called "productionism." In this context I shall refer to this norm simply as "productivity," which I define as a collective value judgment for ranking diverse ways of manipulating organisms to increase yields, to decrease losses in yield and to increase production revenue relative to input costs.

Despite its broad justifiability, productivity can be qualified by other values. In recent decades, agricultural scientists' seemingly unilateral emphasis upon productivity has been criticized on at least three broad fronts. One is environmental quality. Agricultural techniques that increase productivity may have unintended harmful impacts upon environmental quality. A second is international justice. The ethical rationale for increasing productivity is that it advances the well-being of the majority, but some have questioned whether this rationale holds true in international assistance contexts. The third criticism originates from an alleged trade-off between increases in productivity and a decline in the viability of small- to medium-sized family farms in the United States. Here, critics argue that preservation of family farms is more important than increasing net farm productivity. Essays collected by Dahlberg (1986) and by Thompson and Stout (1991) illustrate this point.

I shall not assess the merits of these criticisms here. I mention them because I want to point out the dialectical opposition that arises in each case. Productivity is seen as a valid but problematic thesis guiding agricultural research. In each case, an antithetical value judgment opposes the value of productivity and, by implication, agricultural research as such. Agricultural researchers are faced with a choice. Either they reject the arguments of the critics and defend productivity against the values that stand in contradiction to it, or they must find a synthesis, some way of accommodating the opposing values within the framework of agricultural research. The first of these alternatives may be the one that has been chosen most often. It requires a defense of productivity that either denies the alleged link between agricultural technology and its unwanted consequences, or defends productivity and economic growth in the conventional terms of political philosophy. It is the possibility of opting for synthesis, however, that is of interest here, for this alternative requires scientists to internalize new goals, or, at least, to qualify research-defining values in ways that change the very concept of their disciplinary activity.

The problem now seems to be that while the scientific staff of agricultural research organizations have the capacity to do work that addresses productivity, they lack the capacity to propose or conduct research that addresses environmental quality and other goals. Scientific administration is largely a task of coordinating individual research efforts, so the capacities of existing staff place severe short-term constraints upon choice of research problems. How can this lack of capacity be understood as a defect of philosophical values?

Reductionism

Critics of productivity have traced constraints on research choices to what they call reductionism in agricultural science. Miguel Altieri (1987) and Baird Callicott (1990), for example, have argued that mechanistic reductionism, or "physics envy," in biology represents an explanatory program incapable of recognizing (much less modeling) systems level interactions crucial for an ecological understanding of agriculture. Billie DeWalt (1988) has made a similar criticism of reductionist paradigms in agricultural science but uses the argument to account for inattention to social and cultural dimensions of agricultural production. To be sure, undue emulation of physics may occur in applied biology, but it is a non sequitur to attribute the neglect of alternative research priorities to this form of reductionism. In the present context, reductionism must be interpreted to mean the strategy of breaking broad research goals (such as productivity or environmental quality) into discrete, individualized research efforts that utilize a specific research capacity tied to a discipline, a lab, or, perhaps, to an individual scientist. While this idea was clearly a part of Descartes "method" for the sciences, it is far from the emphasis upon reducibility to physics that is sometimes the focus of the reductionism debate in philosophy of science.

It is useful to think of applied science reductionism as a function of the research capacity implicit in the existence of a laboratory. The laboratory includes the equipment, materials and personnel required to conduct experiments of a certain type. The idea of the laboratory can be used to refer generally to the relevant research unit, whether it be a single scientist at a desk or a research team spread across several locations. There is certainly a range of experiments that can be conducted in any given lab, but the tendency of modern science is for this range to be comparatively narrow. Range is restricted by material equipment and by the technical and disciplinary expertise of the scientific staff. Technology and expertise are also what gives the lab its capacity, so we should expect some trade-off between capacity and range (Latour 1988). A major component of capacity resides in the laboratory's power to control events within its confines. Both technology and expertise help control extrinsic events, so that only correlations of specific interest can account for what happens in the lab.

The applicability of agricultural research depends partially upon a farmer's ability to replicate, or at least approximate, some of the controlled conditions that exist in a laboratory. Although the open fields which serve as labs for some forms of agricultural research are highly uncontrolled when compared to research in the basic sciences, they are far more controlled than a real farmer's fields. Social variables and human factors, in particular, are highly controlled, and environmental variables, though not controlled in a strict sense, may be very unrepresentative of the farmer's field. Even so, the control that exists in an agricultural scientific laboratory may be quite conducive to certain kinds of productivity-enhancing research. To the extent that increases in yield can be defined in physiological terms, for example, the laboratory can control extraneous factors in a search for the biological factors that correlate to increased yield. These results can be transferred to the farmers' fields by transforming the factors that give scientists control over the lab into factors that give farmers more control over their fields.

While new genetic materials and other technologies can transfer some of the conditions of laboratory control to the farmer's fields, the control that a biological researcher has over social variables is not transferable. It is a control that would reside in an absolute power over resources which the farmer simply does not possess. The scientist does not have to make a profit, and if there is an argument with the kids, it does not threaten the future of the lab. The scientist "controls" certain environmental variables, such as long-term accumulation of pollutants, by restricting the time frame of a research project; but this kind of control is not meaningful to farmers, or to those who bear the costs of environmental externalities. Reductionism, thus, influences research in that the scientific capacity inherent in a given laboratory may restrict inquiry into the variables that would be most crucial to research in pursuit of alternative values. To the extent that agricultural research institutions are heavily invested in laboratories that restrict the influence of environmental or social variables, they are limited in their capacity to do research that would serve social and environmental values.

Farming Systems as an Alternative Framework for Agricultural Research

Critics who feel that agricultural science neglects alternative values because of its reductionist research techniques have sometimes called for "holistic" research. Holism is presumed to be the opposite of reductionism, so if reductionism is the problem, holism must be the answer! Holism fails as a theory of research choice, however, because it seems to say "Do everything, and do it now." Even if we could make clear choices in light of the holistic view, holistic management of reductionist research programs is hardly a solution to the problem of revising research agendas in light of existing capacity. If holism is to be an antidote for reductionism, then holism must be a component of the actual research methodology that a scientist employs.

Farming systems research is one name that has been given to research techniques that attempt to resist the evils of reductionism in conducting agricultural research. Some farming systems work makes conceptual links with general systems theory, which strives for holism by "always treat[ing] systems as integrated wholes of their subsidiary components and never as the mechanistic aggregate of parts in isolable causal relations" (Laszlo 1972, 14-15), but the main methodological innovation of farming systems research is to form a team of researchers that interact extensively with farmers. This team is to initiate their research activity with no preconceptions about what the farmer's problems are. By bringing several disciplinary perspectives together in the team's interaction with the farmer, farming systems is expected to be more sensitive to the range of forces that create problems for the farmer. The forces that impinge upon the farmer's activity may not even be primarily agronomic in origin. They may involve family problems or conflicts with neighbors. They may involve access to credit or regulatory restrictions. The inclusion of social scientists in the farming systems team is supposed to help the team listen more attentively to the farmer's problems and to be better able to direct their research capacities toward interventions that address the farmer's needs (Flora and Tomecek 1986).

Any farming systems team will itself have the same sort of intrinsic capacity that has been associated with the idea of a laboratory. The farming systems team is itself a kind of mobile laboratory, consisting primarily in the multidisciplinary expertise of team members. A team with a great deal of expertise may be able to apply agronomic research skills to problems that address family, community or credit problems, but a team that lacks subtlety and imagination may be no better prepared to deal with these problems than the conventional extension service/experiment station partnership that characterizes traditional agricultural research. Indeed, it seems that the "holism" of farming systems research comes down to three things: (1) a recognition of the way that existing laboratory capacities limit the applicability of research to alternative values; (2) an open-mindedness that includes a willingness to listen to practitioners and to researchers from other disciplines; and (3) recruitment of uniquely talented individuals. While these points may be extremely important for the planning and conduct of research, they do not appear to be a particularly deep response to the problem of "reductionism," nor do they require philosophical allegiance to general systems theory. The third point in particular raises the possibility that successful farming systems research depends less upon method than upon serendipity.

If this characterization of farming systems is correct, its advantage over traditional research methodology arises from its practitioner's acceptance of two tenets from practical philosophy and not from deep metaphysical or epistemological views about reductionism and scientific truth. A group of researchers that lacks appreciation of the limiting effects of scientific capacity, or that lacks sufficient open-mindedness, does not itself have the capacity to do farming systems research. But being open-minded and appreciating ones own limitations do not, in themselves, entail that the farming systems team will have the additional capacity needed to formulate agricultural research programs that are applicable to alternative values. Farming systems researchers may accept a commitment to open-ended problem solving, but open-mindedness is far short of a capacity to solve problems related to environmental quality, to small-farm survivability, or to distributive justice. Like productivity itself, these goals must be translated into a technological form if they are to be become the focus of an applied science. Even when these alternative values emerge as clear features of the problematic situation, farming systems teams can direct research toward these goals only to the extent that the researchers have abilities that are not, in themselves, requisite to farming systems techniques (and, indeed, may not even be capable of translation into technological form).

Farming systems may be an antidote to reductionism in the sense that multidisciplinary teams can approach problems with a more comprehensive model of the physical and social reality in which farmers operate. There is a higher probability that the proximate cause of a farmer's problems will be included as a dependent variable in the farming systems model. There is nothing in farming systems, however, that requires or even allows the scientists to bring research techniques to bear upon broader or alternative values. Indeed, the most coherent way to arrive at a holistic model of the farm is to see it as a production system. There *are* alternatives. The farms might be a system for preserving family identity across generations. It might be a system for producing good work habits and a sense of responsibility to others. It might be a feedback loop in a larger system that ensures that human populations do not exceed biological carrying capacity. It would be possible to do research on how agronomic and husbandry practices do and do not further these goals. Indeed, some farming systems teams will successfully address such goals, but they will do so in spite of, not because of, their training in applied science.

Technological Values, Productivity, and Agricultural Science

Busch and Lacy have done survey research in which the various forms of peer review emerge as leading constraints upon agricultural scientists ability to alter the research agenda in favor of alternative values. Patterns of journal publication, peer review of research proposals, and the tenure and promotion process influence research choice and serve as the main vehicles for objective evaluation of research quality (Busch and Lacy 1983). The picture that emerges is a buddy system in which membership in the club requires fealty to existing prejudices. Dundon has offered case studies of agricultural scientists who challenged establishment emphasis upon productivity that substantially confirm this picture (Dundon 1986, 39-51). While I have concentrated on agricultural research, many of these points may be generalizable to medicine and engineering.

In the political economy approach, scientist's research goals define a set of competing interests that merge into the polity of science, an interest which is itself but one interest among others. In political economy, it becomes difficult to ask why individual actors are committed to interests, simply because actors are defined in terms of those interests. The possibility of change or synthesis in the defining goals of agricultural research requires individual scientists to change their allegiance from one interest group to another. The resources for bringing about such a change are essentially two: rational persuasion and brute power.

The analysis that has been given here suggests a modification of the political economy described by Busch, Lacy and Dundon. Scientists in applied disciplines are not, in an important sense, free to change their allegiances in response to either rational persuasion or brute power. The technological character of applied disciplines entails that what these disciplines are primarily about is the discovery of means to achieve certain ends. Scientists are equipped for this task in a dual sense. First, their laboratories are literally equipped (and the research staff is trained) to produce discoveries of a certain sort. Second, the discoveries they can and do produce are made legitimate and important because of their contribution to the achievement of the ends to which they are means. Technological values such as productivity become part of the equipment of a research lab in the applied sciences in the second sense. Changing to new goals is constrained in the same fashion as changing an assembly line from guns to plowshares. First, the actual machines and skills must be changed, for the technology for one task is not the technology for another. The second-level change in technological values is more difficult, however.

Changing an automobile assembly plant into a center for electronic communications might be a more apt analogy. It may appear at first glance that what's needed is a change in material technology, rather than technological values. The facility changes from one organized around the goal of moving people to things to one of moving things to people. One imagines that little more than the building shell will remain should such a change actually take place. Not only the machines, but the people will be replaced. Changing material technology is no more adequate to reorient research than to reorient a manufacturing organization. In point of fact, the past decade was a time in which material technology in agricultural research has undergone a profound transformation. The introduction of gene transfer technologies has completely transformed the material technology of agricultural laboratories in virtually every discipline. What has survived is the idea that agricultural disciplines are organized around the conceptual paradigm of solving production problems. It is only insofar as environmental or health impacts have been capable of being portrayed as production costs that agricultural scientists have been able to respond to the interests aligned in support of these goals. Agricultural scientists just do not know "how to do it" with respect to ecology, sustainability, distributive justice or broad scale rural development. This is a situation that might be changed in the future, but the technological values that make the applied science laboratory what it is cannot be changed without substantial reconceptualization of the disciplinary foundations of the given science. What is more, it is not clear that values such as environmental quality and small-farm survivability, let alone distributive justice, are amenable to the practical rationality that can be employed in pursuit of productive efficiency or human health. It is, in part, because contemporary concepts of practical reason present robust, means-end models attached to the goals of food availability and health that we have developed applied sciences around them. The technological character, in other words, of the foundational values is what makes them work as applied goals for researchers in medicine, agriculture and engineering.

An applied scientist is *not* simply a scientist given a specific practical goal to which science can be applied. If this were so in any genuine sense, organizations dedicated to applied science and engineering would serve a much broader array of social goals than they currently do. An applied scientist gets no credit for having done scientific work when the social problems that get solved are of the wrong sort. Elaborate organizational structures have been constructed to decide and enforce the judgment of which problems *are* the right sort. Perhaps this is a good thing. I have said nothing which would seriously challenge the notion that the applied sciences are appropriately configured as they now stand; but the configuration of these sciences is a philosophical issue that deserves consideration in light of ethical standards and the requirements of justice. Without such consideration, we are technologically committed to means that determine our achievable goals, regardless of the goal's relative importance or defensibility.

The Epistemic and Geographical Boundaries of the Laboratory: An Epilogue

Commenting on an earlier version of the argument in this chapter in 1995, M. Heyboer took me to believe that the role of social factors in changing the direction of applied science research is severely limited. Heyboer took me to task for ignoring "the fact that technological values, like all values, are socially embedded and socially constructed and reconstructed. Productivity means different things in different contexts, and thus motivates different kinds of research in those different contexts." (Heyboer 1995, p. 154) Heyboer goes on to accuse me of constructing an implicit hierarchy of values, with technological values at the top, and with other values subordinated to them. Heyboer takes this to be an attempt to limit the extent to which democratic control over science is possible, and in turn to restrict the extent to which scientists can be held accountable for the effects of their research.

While Heyboer's comments raise a number of legitimate points about the relationship between the argument I have offered here and the emerging political economy/social constructionist school of thought in the philosophy of science, I confess to being a bit mystified at the reaction that the original version of this chapter received. I think of myself as a cautious proponent of that school, rather than an opponent, and would refer readers to *Beyond the Large Farm: Ethics and Research Goals for Agriculture* (Thompson and Stout 1991) and to *The Spirit of the Soil: Agriculture and Environmental Ethics* (Routledge 1995b) for more complete discussions of my views on the philosophy of applied science. For the time being, responding to Heyboer's comments provides a good opportunity to clarify the original paper and to position it within a network.

The key point, I think, regards my use of an inside/outside distinction, the misreading of which has led Heyboer to critique a view I do not hold. It is fair to say that until recently, philosophy and sociology of science have presumed the existence of abstract and indefinite but authoritative epistemological boundary rules that "demarcate" science proper from non-science and extra-scientific activity. Although demarcation was always viewed as problematic, the practitioners of this old-style philosophy of science generally took scientists' links to defense ministries, pharmaceutical companies, regulatory bodies, members of Congress, the Rockefeller Foundation, the World Bank, the AMA, and so on as extra-scientific. Thus emerged an interpretation of "inside" and "outside" that sought to isolate or distill the essence of science proper (the inside) by marginalizing, excluding and ultimately denying the relevance of any environment beyond this epistemological skin. Clearly one implication of Busch and Lacy's work was to reject the implicit hierarchy of so-called inside or epistemic criteria over social networks and the influence of power.

The interpretation of inside and outside that I offer is far more prosaic, however. The inside/outside borders to which I refer are those that someone untrained in science, philosophy or sociology would recognize. Laboratories are places, usually enclosed by bricks and mortar, though sometimes by glass or barbed wire, and my sense of when one is in or out of them relies on the determination that any child would make. This may not be a strictly literal reading of "inside" and "outside," but people can tell when they are in the lab and when they are out of the lab and in the hall, the restroom or the ballroom of the Palmer House Hotel. Clearly a great deal of what both purists and political economists want to classify as science goes on in these other places. Far from wishing to advocate an acontextual philosophy of science, I wish to examine specific contexts, not as theoretical abstractions, but as actual places. I submit that political economy must heed the lessons of political geography.

I *will* defend the claim that what goes on in the places we call applied science laboratories is dominated by technological values, but I do not think that this claim entails that science (even applied science) is always or even usually so dominated in total. I fully accept the suggestion that a complete normative theory of research choice requires the political economy approach, and indeed I have presupposed such an approach even in my earliest work. As such, I reject the claim that my argument "ignores the fact that technological values, like all values, are themselves socially embedded and socially constructed and reconstructed." The technological values in applied science laboratories are technological because they order activity so as to develop products that succeed or fail to the extent that they produce results characterized by instrumental rationality: a cure for a disease, a material meeting certain specifications, or a plant that produces all of its fruit within a narrow time range. They are values, however, because the laboratory embodies a network of social relationships that extends well beyond its walls.

One can (and many of us do) conduct an ethics seminar inside a laboratory, just as one might (and many scientists do) negotiate a contract, write a manuscript or play bridge there. Technological values dominate these other uses of laboratories for three mutually reinforcing reasons. First, the equipment and human capability inside laboratories can be brought to bear upon the production of novel tools. While new tools or solutions to problems might be worked out in many places, laboratories have a geographical advantage over most alternatives that is attributable to the spatial contiguity of equipment and human capability. Of course, this advantage exists only for a limited range of problems that will be specific to each laboratory; that is one of the main points I wanted to establish in the original paper. Second, the people who control access to and from applied science laboratories (in the sense that I control access to my office or my home) are committed to technological values; they generally believe them to be important and worthwhile. At the least their incentives and opportunities are closely tied to the production of tools and the solutions of problems. Third, the enormous capital investment that is required to equip and staff laboratories makes the people that control capital loathe to abandon functional laboratories. To the extent that the administrators and agencies who supply funds think that it is important to get a return on the capital already frozen in a laboratory's technological capability, resources continue to flow into existing laboratories, even when their products are open to criticism.

Each of these sources of dominance, and especially the last two, are social. Technological values have tremendous influence over research choice, but the source of their power does not originate inside the lab. Many of the forces that determine what, if anything, happens inside laboratories are spatially assembled outside the laboratories. Once these forces coalesce in building and staffing an actual laboratory, however, the technological values that dominate the laboratory will resist their realignment. I would accept that this implies a constraint on the extent to which research choice can be made more democratic. It also implies that we should hardly expect a virologist to transform her laboratory into a space where research on attitudes toward sexuality is conducted, even when the virologist in question becomes convinced that the latter is more vital to suppression of the disease that the lab has been organized and funded to control. The virologist should not be held to account for continuing to do work in virology, in other words. This *is* a limit on the accountability of the applied scientist, as Heyboer suggests.

But technological values limit the broader accountability of applied scientists far less than they limit the ability of applied scientists to change what they do inside the laboratory. I agree with the claim that applied scientists can and should question the normative implications of their practice. Whether this questioning must take place inside the laboratory is a different and more difficult question to resolve. Certainly it will not be adequate to confine ethical deliberation to laboratories, since the walls of the laboratory will then preclude precisely the sort of public, democratic discussions that are necessary for changing the ensemble of forces that cause laboratories to be constructed according to a given regime of technological values, in the first place. Heyboer's comments, thus, usefully bring out the implications of my analysis for the accountability of scientists, but that was not the main point of my argument.

I wrote the first draft of this chapter in 1989. It was focused more narrowly on agricultural science and was presented at the AAAS meetings in 1990. At that time, agricultural science administrators had not completed (though they were well into) a massive investment in molecular biology laboratories. The original paper was intended to reflect upon an earlier generation of administrators' investment in laboratories that were incapable of responding to critics of chemical and mechanical technologies for agricultural production. Those new laboratories have now been built and staffed, and they will either determine what agricultural experiment stations do for the next few decades, or the enormous financial, political and emotional investment in their construction will simply have to be abandoned. One role for normative philosophy of science is to develop the argument for abandonment. Another is to undertake questioning and deliberation within the constraints of this investment in technological values, but the latter task requires that we understand the constraints, partly by asking how (and if) they function in other areas of applied science. That is what I take this chapter to be about.

Part Two

Teaching

4

Undergraduate Education for Agriculture and Natural Resource Professionals: Emphasizing the Social Sciences and Humanities

Many authors writing on the future of undergraduate education in colleges that have historically had their roots in agriculture have stressed the need for a broad view of society and culture. Virtually no one has called for more specialized training in the applied sciences. What is more, the completion of the Social Science Agricultural Agenda Project has produced a wealth of material for those who wish to find more detailed discussion of the topic (Johnson and Bonnen, 1991). Three key points need to be made with respect to the role of the social sciences and humanities in the education of agricultural and natural resource professionals.

1. Social science and humanities courses play a dual role in undergraduate education. They are a part of core undergraduate education, but certain social science and humanities topics have special relevance for the careers that agriculture and natural resource professionals will pursue in the twenty-first century.

2. The social sciences and humanities are the only disciplines within the university that are equipped to help students understand the way that an increasingly urban population will perceive food and environmental issues. If agriculture and natural resource professionals are to be effective in their careers, they must be prepared to listen and reply to concerns and desires voiced by people with little life experience or formal education in the production of food and fiber or in the management of natural resources.

3. The capacity for targeted social science and humanities education on topics and skills crucial to future professionals in agriculture and natural resources is both low and poorly organized. This area has been among the most neglected by faculty administrators in both agriculture and environmental sciences and in the liberal arts.

The balance of this chapter takes up each of these three points in turn. The importance of a broad education is taken as a given; there is no further discussion of it here.

The Difference Between Core and Targeted Education in the Social Sciences and Humanities

Many of the contributors to the National Research Council publication where this chapter was originally published cited the need for broadening the curriculum of agriculture and natural resource professionals. They cited the need for courses in communication, foreign language, and in a core area of knowledge about culture and society. Yet there was nothing special in their arguments that relates to agriculture or natural resources. Business students, engineering students, premedicine or prelaw students have the same needs. It is a real need that must be acknowledged, but recognition of this need should not influence the curriculum reform effort in agriculture and natural resources to focus only on developing a core curriculum.

Agriculture and natural resource professionals face special problems of communication, ethical decision making, and of interpreting and managing human activities—problems that are unique to the type of careers they will follow and to the kind of science they will apply. The social sciences and humanities can and must be incorporated into their training in such a way as to target educational efforts on the acquisition of knowledge and skill that is specifically relevant to these problems. Some of the specific topics that should be targeted are discussed below (as well as in other chapters).

The single most important point that must be recognized, however, is that emphasis upon core humanities and social science does not substitute for the targeted education on special social science and humanities topics of particular importance to agriculture and natural resources. Programs of core education that stress "great books" or a unified approach to understanding society and civilization through the study of art, literature or history are quite likely to *exclude* these special topics in a systematic and deliberate way. Put simply, agricultural ethics, communication and even history are at best tentatively embraced by the purists of the basic disciplines. If these topics are not introduced into the education of agriculture and natural resource students at the upper division or graduate level, the core education movement to emphasize the social sciences and humanities will, in fact, deprive future professionals of the social science and humanities knowledge that is most crucial to their effectiveness. To note this is not to oppose the core movement, for all students need it. It is crucial to see that there are two tracks for discussing the role of social sciences and humanities in the education of agriculture and natural resource professionals. One is the core needed by everyone, the other is the targeted areas needed for agriculture and natural resources. Everyone needs the core. Agriculture and natural resource students need something else, preferably in addition to the core. What it is that they need is the topic that follows.

The Content of Targeted Areas in Social Science and the Humanities

American agriculture enjoyed a reputation for success during much of its tenure, but the most recent decade has been one of criticism, and rethinking of agricultural priorities (Johnson 1984a; Danbom 1986; Kirkendall 1987). One important group of critics, associated with 1972's "Pound" Committee of the National Research Council (NRC) (NRC, 1972; 1975), stressed the scientific quality and efficiency of agricultural research, but the more noted critics have focused upon the social goals that contemporary agricultural production techniques (whether implicitly or intentionally) have tended to serve (Hightower 1975a; Berry 1977; Jackson 1980; Schell 1984; Doyle 1985; Fox 1986). There is an extraordinary range of concerns and complaints expressed in the writings of this latter group of critics, and many different client groups are alleged to have been ill-served. A theme common to most criticisms, however, is that agricultural leaders have, *de facto* or by design, pursued a goal of maximizing the productive efficiency of the American farm. Critics allege that it is the USDA/land-grant system's persistent search for greater yields that is the wellspring of problems for American agriculture. The tomato harvester and BST cases discussed in chapters 1 and 2 are but two of many that complicate the current context of agricultural production, research, distribution, and consumption. Others include the question of field testing engineered organisms, determining acceptable risks associated with agricultural chemical residues in food, examining our commitment to the development of Third World agriculture in light of domestic farm interests, the preservation of genetic diversity, and the question of environmental and public health regulations as barriers to trade in agricultural products. These are among the most difficult of a long list of topics that, more broadly, include world hunger, environmental quality, animal welfare, and the traditional agrarian philosophy of farming as a way of life.

Today's agricultural leaders, not to mention today's citizens, need a more sophisticated understanding of the food system. They need to appreciate the social, ethical and cultural values that seem to surface only in a crisis situation. While it is not clear that crises such as the banning of Alar or California's "Big Green" referendum can be anticipated with any degree of confidence, a deeper understanding of the social and cultural forces that are operative during crisis situations will help agricultural leaders make more effective and responsible responses to public concerns, when they arise.

Traditional agricultural education in the humanities and social science has stressed economic management of farms and agribusinesses, as well as rural community development. While there will be a continuing need for this education for a percentage of students being educated in traditional agricultural programs, there is an even greater need for education on how society beyond the farm sector relates to and perceives agriculture. Future professionals will need to know how to do a better job of producing products that urban consumers want. They will need to know how to manage their professional responsibilities in a manner consistent with the public interest. They will need to know why people who do not have a farm or rural background might find certain practices or ways of speaking arrogant, insensitive or otherwise objectionable. They will need to know how to listen to a new constituency.

Present day farm, food industry and environmental leaders are keenly aware of the fact that the consumer and political decisions that define the framework for agriculture and resource management are made by people who have no life experience or formal education in agriculture or in environmental science. The general public did not grow up on farms and have little or no experience with the productive use of natural resources. Many lack life experience even with recreational use of nature. The dietary choices and opinions on regulatory issues of most Americans are not informed by knowledge of principles for evaluating and comparing risks, nor by information on the contribution of existing farming and management practices to food availability, economic growth, and the provision of other human needs. While we should work to promote better public understanding of agriculture and resource management, we must plan the education of the next generation of leaders on the assumption that this situation is not likely to improve. It would be tragically foolish to think that the public at large will assume the personal costs needed to understand the scientific- and production-based opinions of those who make their lives in agriculture, the food industry and resource management. It is agriculture and natural resource professionals who must bear the responsibility to communicate with the public. This means that leaders must understand public opinion on its own terms. Put simply, the mountain will not come to us; we must go to the mountain.

Specifically, this means that undergraduates need to study how nature and natural resources are perceived in American society. They need to learn alternative philosophical approaches to the measurement and acceptability of risk. They need to be taught how scientific advances such as biotechnology are received by different sectors of the public, and whether public reactions are based upon political and financial interests or upon moral and religious concerns. There is a need for graduate and undergraduate training in communication strategies that do not alienate non-scientific, non-farm audiences. There should be undergraduate courses that take up the politics of the policy process, and the history and organizational structure of groups (commodity organizations, environmental or animal welfare activist organizations, consumer groups, etc.) that influence agriculture and natural resource practice. Journalism departments should offer courses on how news media decide which stories to cover, and of the norms and institutions that structure the coverage of science and technology. Future professionals who possess knowledge and skills in each of these areas will be far more effective in conducting and promoting research, product development, marketing, management and policy change than those who do not. The volumes of the Social Science Agricultural Agenda Project (Johnson and Bonnen 1991) document the subject matter for such targeted course work in an exhaustive manner.

Yet it is clear that the current social science capacities within colleges that have their roots in agriculture have little capacity to do undergraduate education (much less research and extension) on these issues. This lack of capacity is partly organizational. Marketing and resource economics courses in agricultural economics are targeted for advanced majors. A similar situation holds for course work on social psychology, development theory and cultural analysis that might be offered by sociologists and anthropologists with agriculture and natural resource appointments. Such organization does little to serve the broad educational needs of undergraduates. A more serious problem exists with respect to educational needs that derive from political science, communications, journalism, literature and philosophy. With the exception of agricultural communications or journalism programs aimed primarily at training tomorrow's agricultural press, capacities for educating undergraduates on these topics within colleges of agriculture and natural resources are practically non-existent. What, then, are the barriers to reform?

Barriers to Emphasizing the Social Sciences and Humanities

During the 1980s, a number of journals and professional societies emerged to support research on the broad social and ethical issues that spawned conflict, controversy and the need for better communication between agriculture and natural resource professionals and the public at large. *Agriculture and Human Values, Issues in Science and Technology,* and *The Journal of Agricultural and Environmental Ethics* have joined traditional outlets such as *Science, Environment,* plus other environmental journals and many monographs and anthologies devoted to agriculture and natural resource issues. The Agriculture, Food and Human Values Society had over 700 members by 1991. The basic knowledge for meeting teaching needs exists to a far greater extent than it did in 1980. Although there will be a continuing need for research and publication in the areas of agriculture, environment and societal values, lack of models and materials can no longer be accepted as an excuse for not offering educational opportunities to undergraduates.

Curriculum reform is a faculty-based process: decisions about what is to be taught are ultimately made by faculty. Since faculty are not likely to place themselves in a position in which they are expected to teach subjects and methods in which they perceive themselves to have little expertise, the existing capacity of agricultural faculty places a severe constraint upon the direction and extent of change in the agricultural curriculum. Curriculum reform has largely meant that standard courses in the plant and animal sciences substitute the study of gene transfer and computer technology for the study of mechanical and chemical technology (and it is a partial substitution, at that). In some instances, courses that approach production problems in terms of cropping systems or farm management are being replaced by course work that takes an even narrower approach, generally assuming that an ability to identify the genetic basis of economically valuable traits need be the only item in agricultural scientists' tool kit for the coming generation. Even in the social sciences, the response has often favored replicating management and computer systems curricula currently offered in colleges of business.

To the extent that agricultural curriculum efforts have tended to increase capacity for exploiting discoveries in molecular biology and computer technology, one can argue that they have failed to respond to the needs outlined in this chapter, entirely, and may, in fact, constitute an abandonment of the special historical mission of agricultural education. The emphasis upon technology responds to declining enrollments by introducing a curriculum that undergraduates perceive to offer training in marketable skills. Biotechnology and computers are not peculiarly suited to agriculture and natural resource management, however, and the new corps of undergraduates correctly perceive that their ability to exploit their technical training in no way depends upon a sophisticated or reflective understanding of the food and resource system. The state of agricultural education's ability to investigate and disseminate a comprehensive and unified vision of food systems in modern society has, if anything, been damaged, rather than improved, by curriculum reform efforts that stress hot new technologies.

Faculty may have also thought that such a stress would be consistent with existing research and educational capacities of agriculture and natural resource faculty. In fact, however, the move has been accomplished by importing new faculty whose training and experience gave them no particular basis for loyalty to agriculture or resource management. In effect, the move has allowed the administrative structure and faculty lines of formerly agricultural colleges to be captured by both students and young faculty who have no particular interest in or understanding of agriculture, natural resources, or of the social systems that support them. This capture is, of course, partial; a majority of faculty and administrators still have traditional roots in agriculture and natural resources. However, the fact of this capture points us toward both problems and opportunities for agricultural colleges to respond to the need for change in their curricula. New faculty (and students) will not only lack the capacity to deal with a broader notion of agriculture, they will also lack any reason to regard their educational mission as encompassing such broader concepts. It may soon be clear that enhancements in the direction of biotechnology and computers respond to declining undergraduate enrollments at the expense of the traditional land-grant mission, and that they leave food producers, resource managers, and rural communities without any educational organizations that are committed to the creation and dissemination of knowledge in support of their interests and ways of life.

On the other side of the equation, there is reason to doubt that the liberal arts disciplines can supply teaching expertise in the required areas on many land-grant campuses. Although there are many individuals that have such expertise, they are not equally distributed throughout all universities. Liberal arts departments that have concentrated on achieving disciplinary expertise or quantitative skills in areas such as sociology, political science, history, philosophy and literature are quite unlikely to have hired and promoted faculty members who specialize in science policy, environmental studies, risk issues, or science communication during the past decade. Simply inviting these departments to offer course work for agriculture and natural resource students would produce a disaster at some universities.

There are, of course, more familiar and mundane barriers to the advancement of an agricultural education system with a broad and sophisticated vision of the social, cultural and ethical dimensions of the production and distribution processes, consumption patterns and management possibilities for renewable resources. Tenure and promotion, opportunities for publication, and funding sources all come readily to mind. To a large extent, however, these more commonly cited barriers are all functions of the existing research and educational capacity within agricultural universities, since each is the result of expectations and values that are held by individuals who currently occupy faculty and administrative posts. While it would be naive to ignore such barriers in promoting curriculum change, it would be equally naive to think that an effective effort to enhance agricultural faculty's ability to integrate values issues into more technical subjects will not simultaneously improve the prospects for overcoming institutional barriers of this sort.

Opportunities for Change

There are also a number of institutions where innovation in the social sciences and humanities has been successfully targeted to subject matter areas needed for agriculture and natural resource professionals. In each of the efforts where enhancement has succeeded, a four-stage process has been followed. First, there has been a careful effort of planning and coordinating a "core group" drawn from a variety of disciplines. The members of this group must agree to work together over a long period of time and must be willing to respect the integrity of other members. This is especially the case when the group must agree to disagree on some point, but must get on with the business of planning and coordinating larger activities. Second, there has been explicit attention to what might be called "market development," for the activity that is to be carried out. In the case of a group that produced a book on research policy at Texas A&M (Thompson and Stout, 1991), this consisted of identifying faculty and administrators who would take time to participate in some of the workshops offered by the group. The third stage is the dissemination of information through workshops or conferences or workshops that, by outward appearance, resemble conventional academic activity. This is, frankly, the most efficient way to disseminate ideas among a faculty that is used to the idea of attending conferences. Finally, there has been a follow-up phase in which faculty from the various disciplines maintain contact, sometimes in a systematic way by initiating more structured collaborative projects, but often by way of informal networking. These four stages are not necessarily a temporal succession; they represent levels of activity that can be pursued simultaneously. The possibility of enhancing capacity on ethics, social values and food systems depends upon understanding how each stage presents tasks that must be accomplished if the goal is to be achieved.

Planning. All activities require planning, of course. What is special here is that the planning group includes people who must talk across disciplines to one another, and who will be sensitive to the barriers that are imposed by jargon and by the reigning values of people in different academic departments. This group must develop a rapport and must be willing to make a multi-year commitment, though the total number of hours required from each participant may be quite small.

Marketing. The effort required to find and prepare an audience for the primary product will depend upon many decisions that are made in the planning process. At a minimum, the success of the project requires preliminary "networking," identification of agriculture and natural resource faculty members who will be supportive. The process also requires gestures from the administration that demonstrate their seriousness. The creation of positions and the expenditure of money may be painful, but there is no better way to communicate seriousness.

Transfer. This is something that academic professionals know how to do, but organizing, advertising, administering and presenting workshops is both time- and money-consuming. In this area, workshops should stress experiential learning techniques, case studies, and role playing simulations for students. These approaches are essential if faculty who do not have disciplinary expertise in values studies are to teach values issues in the classroom. They are also the most effective educational techniques for students who are pointed toward careers in agriculture and business. Although learning modules that use these techniques must be constantly developed, updated and refined, this is one area where the work of the past decade has put us in good stead to accomplish some dissemination of modules in the next century.

Follow-up. Follow-up includes more networking to assure that those who participate in workshops continue to receive information and support. There should be nationally coordinated efforts so that faculty who are not trained in social science and humanities disciplines have continuing access to those who are. Like marketing, the activities of follow-up depend a great deal upon the circumstances of particular individuals, so much of what must be done cannot be described in advance. There has been too little organized follow-up from previous agriculture and liberal arts projects, as well as from curriculum development activities sponsored by the Office of Higher Education, USDA.

5

Agricultural Ethics: Content and Methods

The previous chapter makes the case for agricultural ethics within the curriculum for higher and secondary education and describes some of the main themes that would be taken up in teaching agricultural ethics. This chapter discusses the actual content of a likely course or unit on agricultural ethics. I developed one of the first college-level courses in agricultural ethics at Texas A&M University and have taught it from 1982 until 1997. This allowed me to experiment with different course delivery methods including case studies, decision cases, role-playing, team-teaching, and distance learning technology. My participation in workshops on teaching agricultural ethics on numerous occasions has also given me an opportunity to receive feedback from others who are teaching agricultural ethics. This chapter is a personal and subjective summary of my fifteen years worth of experience teaching this subject matter.

There are a host of issues that need to be taken up in any decision to offer agricultural ethics to college undergraduates. Not the least of these are the institutional decisions about how to staff and evaluate the instructional effort. In the simplest case an individual faculty member simply decides to begin offering agricultural ethics course work in an existing course, but the simplest case is also probably quite rare. More typically neither agricultural nor philosophy programs are offering courses with natural affinities to agricultural ethics nor are they likely to have individuals who are both competent and interested in offering ethics instruction for agriculture. The course in agricultural ethics at Texas A&M University owes much of its current shape to decisions that were undertaken in 1979 by then Dean of Agriculture H.O. Kunkel and an advisory committee that included Edward Hiler, the Dean and Vice Chancellor for Agriculture and Natural Resources at this writing. This constitutes a fifteen-year history of high-level support for coursework in agricultural ethics that is unparalleled by any other institution, but agricultural ethics would not have become an established course at Texas A&M University were it not for the fact that the original proposal was also warmly received by John J. McDermott, then Head of the Department of Philosophy and Humanities. These and other key figures decided that a faculty line would be committed to agricultural ethics long before I had even completed my program of graduate study.

It has proved difficult to match this level of support and receptivity in many other universities. Where there are strong and long-standing programs in agricultural ethics at other institutions, they tend to be supported and staffed entirely within agricultural programs. Jeffrey Burkhardt has developed a very strong program in agricultural ethics at the University of Florida, but his links to his disciplinary home in philosophy are wholly informal. The first course in agricultural ethics was team taught by agronomist Thomas Rheur and philosopher Stan Dundin at Cal-Poly, San Luis Obispo in 1980, but within a few years only Rheur was receiving support to teach it. There are also cases where willing philosophers do not find themselves in an institutional setting that is conducive to teaching agricultural ethics. Some of the best known philosophers who wrote on agricultural ethics from 1975 to 1995—Gary Comstock at Iowa State University, Charles Blatz at the University of Toledo or William Aiken at Chatham College—had few opportunities to teach courses in their areas of expertise.

Although institutional decisions shape the content and methods for teaching to a considerable degree, the often local and circumstantial nature of institutions makes this topic unpromising for the present chapter. Of wider interest is the matter of what to do once one finds oneself in a position to do something. Almost any college level instructor *can* incorporate at least a module on ethics into an existing course on agriculture. The modular approach thus represents one model—albeit a minimal one—for teaching agricultural ethics. The full semester course taught by a specialist—the Texas A&M model—represents the opposite extreme. In between are team taught courses on "issues in agriculture" that might involve both philosophers and agricultural faculty, or that might incorporate guest lectures, either in person or using distance technology. This chapter reviews some alternative visions of what to teach in either a module or a course on agricultural ethics as well as a candid assessment of the potential and limitations of several common delivery systems.

What to Teach

A teacher of agricultural ethics is not a preacher or proselytizer for a particular point of view. Teaching agricultural ethics means giving undergraduates a better understanding of why key issues *are* issues, why people disagree. It also equips them with critical thinking skills and concepts for personal reflection, for deciding what is right and what is wrong on their own. Two thousand years of research on ethics has produced a family of principal ethical theories or general conceptual approaches to understanding why one thing is right, another wrong. An instructor who is literate in ethics will be able to teach students some concepts and patterns of philosophical reasoning that will transcend the particular subject matter taken up in any given course and provide the grounding for life-long learning and reflection on ethical responsibilities not only in agriculture, but in other walks of life. An instructor literate in ethics, however, is not necessarily literate in the problems of agriculture, and vice versa, hence the main source of difficulties in deciding what to teach in agricultural ethics.

This chapter shows that the potential subject matter of agricultural ethics is as broad as agriculture itself. Subject matter decisions will clearly be determined by individual instructors in most cases, and by institutional committees in those instances where an orchestrated decision to offer or accredit coursework must be made. Agricultural ethics includes environmental dimensions of agriculture, the aesthetic, moral and even religious acceptability of food production, processing and consumption practices, the problems of population and hunger, the use (and abuse) of agricultural animals, the organization of agricultural labor, the sustainability of agricultural production, and, of course, the long and complex historical transition from agrarian to industrial society. This is a broad list and the actual turf is even broader. I see no reason to constrain agricultural ethics to some list of "official" problems or subject matter. The specific subjects will be those that most closely fit a matrix determined by course structure, instructor interest and competence, and student demand.

The topic for this section is thus not what to teach in the sense of world hunger or animal rights but in the sense of what kinds of reading materials to assign and how to structure a session on ethics, without regard to subject matter. The approach that most philosophers use is to assign readings on an issue or topic that represent several different philosophical approaches to the topic. The instructor then exposits and analyzes the readings in a lecture/discussion class format, and assigns essays that require students to do the same. The paradigm example of this way of teaching applied ethics is the abortion debate. An extensive literature of opinion and argument on abortion exists, and perspectives on the ethics and politics of abortion often turn upon important differences of philosophy: what it means to be a person, whether to evaluate action solely in light of its consequences, and the like. Undergraduate texts for philosophy often consist of articles that provide the gist for classes that follow this paradigm. Two that are of direct applicability in agricultural ethics are Singer and Regan's *Animal Rights and Human Obligations* and Aiken and LaFollette's *World Hunger and Morality*, both out in succeeding, updated editions since the late 1970s.

Teaching from Anthologies

Charles Blatz's (1991) *Ethics and Agriculture: An Anthology on Current Issues in World Context* is a large collection intended to cover the full range of issues that might be taken up in agricultural ethics. It has clearly been assembled with the implicit paradigm for undergraduate courses in philosophy in mind. However, with all due respect for Blatz's truly impressive editorial achievement, instructors who attempt to use this collection to teach agricultural ethics are likely to experience some frustration. For a few topics in particular, Blatz has found articles that present the contrast between differing philosophical perspectives beautifully. Three articles on water policy are especially useful.[1] For the most part, however, while the articles that Blatz has collected clearly represent different perspectives on agricultural issues, it is far from clear that the difference in perspective can be systematically linked to the kind of conceptual distinctions that characterize distinctions in ethical theory. The upshot is that the Blatz volume is a very unsatisfactory tool for getting elementary ethical distinctions such as the difference between rights-based and consequential approaches across to undergraduates. Yet if one attempts to teach the articles on their own terms, one is too frequently swept immediately into very complex debates that integrate philosophical principles closely with economic, sociological and anthropological analysis. It is ar-

1. They are Scherer's "Toward an Upstream-Downstream Morality for Our Upstream-Downstream World," Anderson and Leal's "Going with the Flow: Expanding the Water Markets," and Ingram, Scaff and Silko's "Replacing Confusion with Equity: Alternatives for Water Policy in the Colorado River Basin," all in Blatz, 1991. I have reviewed all three of these articles, including a discussion of how they illustrate libertarian, utilitarian and egalitarian reasoning in my contribution to Katz and Light's *Environmental Pragmatism*. See Thompson 1996.

guably philosophy of social science, not ethics, that lies at the root of the different opinions held by authors such as Russell and Zerwekh (see Blatz 1991, pp. 618-629), whose analysis of the Green Revolution is sharply critical, and that of Ruttan and Hayami (see Blatz 1991, pp. 637-650) who are less so, but even if a rights/consequences debate is buried here somewhere, it cannot be readily surfaced from the texts themselves. There are clearly important philosophical issues at stake, but it is less clear that they can be readily taught to undergraduates who are only learning the philosophical vocabulary.

My own reading of the literature suggests that Blatz has done about as well as could have been done, however. The problem is that the authors of most of the articles collected in *Ethics and Agriculture* were not generally writing with philosophy and ethics in mind. Indeed, many of these authors probably lack the philosophical vocabulary to have written an ethical analysis themselves. Where a problem has attracted the attention of philosophers (such as the animal issue or world hunger) there is a literature that can be used to teach ethics to undergraduates, but where it has not, the literature can only be discussed philosophically at a level more appropriate for advanced or graduate students. The unilateral experience of agricultural ethics teachers with whom I have contact is that the vast majority of undergraduates being exposed to agricultural ethics materials will not have previous exposure to philosophy or philosophical concepts, hence the prospects for advanced treatment of agricultural ethics are limited.

Unfortunately, even where the work of philosophers is available there are limitations. The animal welfare/animal rights literature is a case in point. Chapter 8 takes up the animal welfare/animal rights distinction in some detail, so the substantive discussion of this issue will not be taken up in this chapter. Philosophers have found writings on the moral status of animals irresistible. Debates over laboratory testing and animal agriculture enliven a philosophical literature where the main points of contention are represented by the points of disagreement between Peter Singer and Tom Regan. Singer and the consequentialists think that human-animal relations need reform because the pain that animals endure at the hands of humans is not balanced by sufficient benefit, while philosophers such as Regan think that reform is needed because human use of animals is inconsistent with respect for cognitive and biological interests that cut across species but are foundational for the entire project of morality. This is an important philosophical difference, but it frustrates many agricultural students (and faculty) to have a treatment of "the animal issue" in which "both" sides are sharply critical of present human practices for using food animals. This is the "apparent bias" problem: from the standpoint of agricultural students, it looks like *all* philosophers are biased against their deeply held views.

There are philosophers that take up alternative points of view, of course. R.G. Frey is perhaps the best known philosophical "opponent" of animal rights, and his work shows up frequently enough in philosophy classes. But Frey's opposition is deeply philosophical. His most famous work, the book *Interests and Rights* takes up the animal issue as a vehicle for a more thorough philosophical study of the link between interests (which animals clearly do have, in Frey's view) and rights (which they may not have). In a recent article, Frey devotes most of his effort to exhibiting why it is more difficult to justify the use of animals for human benefit than many scientists who take up the question have been willing to admit (Frey, 1995). While I read this as a plea for more careful philosophy, others might as easily read it as yet another "pro-animal" philosopher espousing his opinions.

The other key figure in the philosophical literature is Bernard Rollin. Philosophers who know Rollin only through his 1981 book, *Animal Rights and Human Morality,* may view him as a precursor to Regan: advocating an animal rights view, but with less rigor and without the extensive review of the philosophical literature that is found in Regan's 1985 book, *The Case for Animal Rights.* In fact, Rollin's view is quite different from Regan's, both philosophically (it is based on the notion of an implicit social contract with animals) and practically (it permits carefully considered human uses of animals, including food and biomedical research). Rollin has recently produced the most extensive study of farm animal welfare to date, and the book that should be the touchstone for any serious discussion of the issue (see Rollin, 1995). While it remains to be seen whether this new study will crack the Singer/Regan monopoly on philosophical approaches to the animal issue, using Rollin's work in a course on agricultural ethics does not solve the problem of apparent bias, for he, too, is best known by animal scientists as a critic of animal agriculture.

In contrast to any of these philosophers, the advocate of animal agriculture thinks that what should be questioned is the premise that human-animal relations need reform at all, at least with respect to food production. The best known advocate of *this* position is Stanley Curtis (anthologized in Blatz as well as in Singer and Regan). But Curtis is an animal scientist, not a philosopher, and Curtis's position gets little attention from philosophers who are primarily interested in using the animal welfare/animal rights discussion first as a pedagogical tool for illustrating the conceptual difference between welfare and rights. Ironically, Curtis's scientific work has taken a strong turn toward empirical investigation of the features of sentient beings that are crucial to the moral consideration of their interests. This work has received little attention from philosophers, while it has led some animal scientists to perceive Curtis himself as a critic and advocate for animal rights. The upshot is that whatever materials one chooses to use in an agricultural ethics course, students from farm backgrounds feel thoroughly excluded from the debate as configured by philosophers, and this felt exclusion is not a promising starting point for teaching them agricultural ethics.[2]

Teaching from Textbooks

The perception that existing materials were less than ideally suited gave rise to the Ethics Team of the National Agriculture and Natural Resources Curriculum Project (NANRCP) in the 1980s. This project was initiated by the U.S. Department of Agriculture's Office of Higher Education and funded in part by them and by a consortium of other donors, including many private companies such as Upjohn and RJR/Nabisco. In addition to a broad endorsement of the need to improve the teaching of ethics, the NANRCP Ethics Team concluded four years of study on curriculum needs in agricultural and environmental ethics with two recommendations. First, the team advocated the development and use of case materials for teaching agricultural ethics to college undergraduates. Case materials are discussed at some length below. Second, the team recommended the completion of a stand-alone textbook that would have self-contained expositions of ethical concepts, their application to agricultural issues, and overview chapters discussing the main issues. *Ethics, Public Policy and Agriculture,* the book I co-authored with Robert Matthews and Eileen van Ravenswaay, was the result of that recommendation. *Ethics, Public Pol-*

2. This assessment of the prospects for teaching the animal issue in agricultural ethics illustrates the problem with philosophical anthologies, but it may leave a too negative impression of the prospects for teaching the issue to agricultural students. It is true that a significant minority of students seem completely unable to make a rational assessment of philosophical writings on food animals. They see them as utterly inconsistent with what they hold sacred, and they move immediately to the judgment that since these philosophers cannot mean what they say, they must be up to some evil and nefarious purpose that is not revealed in their words. However, even these students seem to enjoy having an opportunity to study the issue in class. What is more, many agricultural students learn a great deal from the animal issue, even if they do not change their views. The issue clearly belongs in an agricultural ethics class, and most students will be receptive to it.

icy and Agriculture begins with three theoretical chapters that describe basic approaches to policy analysis, to ethical theory, and an approach to combining the two. Although the approach to linking ethics and policy analysis is worth discussing in its own right (see Chapter 7 of the present volume) the textbook has proved to be both too much (too complicated) and too little (important ethical and economic concepts omitted) for everyone who has used the book with undergraduates, including myself. The balance of the book consists of chapters on food safety, environmental impact, animal well-being, world hunger, sustainability, and saving the family farm. Though far from complete, this is a representative sample of key issues in agricultural ethics, and it is through these chapters that the book does serve as an alternative to Blatz's anthology approach. Readers who have an abiding interest in "What to teach" are advised to consult *Ethics, Public Policy and Agriculture.* Nevertheless the authors would be the last to suggest that this book is truly "self-contained" as the NANRCP Ethics Team had hoped.

Deciding what to teach is difficult in large part because agricultural ethics has been a neglected field with too few participants. Medical ethics does not suffer this problem, despite the fact that the number of students attending medical schools is comparable to the number of agriculture and natural resource students. Most medical schools have established medical ethics programs, and many of the medical and philosophical faculty in these programs produce publications that are suitable for classroom use as a component of their research. In 1996, fully fifteen years after the experimental courses were offered at Cal-Poly, Texas A&M, and the University of Florida, some of the largest and most prestigious agricultural programs in the world had yet to offer a course, and others were still in the experimental stages. It is not clear that the field will catch on even at this writing, yet if it does, the "What to teach" question will solve itself. Even now work that has been published serials such as *Agriculture and Human Values, The Journal of Agricultural and Environmental Ethics,* or *Choices* could be collected to produce a better anthology. Sooner or later, someone will do so. The absence of institutional support at the level described in Chapter 4 is the reason why finding appropriate course materials can still be frustrating, but the future looks brighter.

Cases

Given the problems that exist with basic text material, one response has been to resort to cases, but there is considerable ambiguity about what constitutes a case. Legal case law provides one model. Here the relevant facts about a lawsuit or criminal trial are summarized, and the judge renders a written opinion citing legal precedents and explaining why the cited precedents are being interpreted as relevant to the case at hand. Legal case law produces these transcripts as part and parcel of the process of judicial review, but they are often excellent teaching tools precisely because the facts of the case are reproduced along with an account of the judge's evaluation, complete with detailed justification. Supreme court cases are especially important because dissenting opinions provide multiple perspectives on the issue at had. The *legal case model* is often simulated specifically for teaching purposes. An author summarizes the facts involved in a given series of events, then provides an analysis of these facts that parallels the role of the judge in a legal case. So "the tomato harvester case" or "the BST case" are analyzed throughout this book. The facts are summarized briefly, and an analysis ensues.

The legal case model is better than a purely abstract discussion of ethical issues for obvious reasons. Concepts are applied to an example and they become concrete. Since the cases are real, students can see both that and how concepts are relevant to real world situations and dilemmas. Such cases are open-ended teaching tools, in that students may augment them with additional information and may build their own analyses. But to be honest, there is nothing pedagogically different about using "cases" in this sense than from using examples to flesh out a lecture. If there is a point to be made it is a theoretical one: some would claim that such cases are the *medium* for ethics itself, just as actual legal cases are the medium for law itself in common law nations such as England and the United States. If this is one's view, then cases have a somewhat different philosophical significance than they do for one who believes that ethical principles have a rational or empirical foundation apart from their application in cases. It is advisable for those who advocate teaching from cases to be aware that their advocacy may be linked to a particular view of ethics, but this philosophical debate does not have further implications for the topic of this chapter.

A *conflict simulation case* attempts to describe the roles and interests of several groups that may have found themselves at odds in a conflict. The Chatham River case from Wilson and Morren's *Systems Approaches for Improving Agriculture and Natural Resource Management* is an excellent example of a conflict simulation case. Three interest groups with competing uses of the fictional Chatham River's water are described. Farmers want the river water to irrigate crops and water livestock. The Town of Springdale wants the water for economic development. The

Friends of the Chatham, an environmental action group, want the water to stay in the river where it supports wildlife and recreation. The most plausible arguments to support each interest replicate libertarian rights-based, utilitarian and egalitarian-ecocentric philosophies, respectively. The case can be used simply to describe and analyze a conflict, or it can be used in a role-playing exercise in which students are assigned to an interest group and told to come up with the "best" argument. Although placing this much control in students hands can be risky, students do in fact tend to come up with libertarian, utilitarian and ecocentric arguments for each interests' position on the use of Chatham River water. When one can elicit relatively pure versions of philosophical arguments from students' own handling of a case, one can make a very convincing demonstration of the point that philosophical differences truly matter in public policy (Thompson, 1996).

Conflict simulation cases can be quite extravagant. Another case developed through a collaborative project between Texas A&M and Penn State simulates a conflict among stockholders of a mass market burger restaurant chain. Some stockholders have taken relatively strong animal welfare and environmentalist views and want the company to change its menu and suppliers in accordance with their vision of "ethical investment." Other roles for the simulation represent both restaurant management and public relations divisions of the corporation, as well as animal commodity organizations and the press. In this case, students are placed in groups that correspond to these roles and must complete outside reading and research to support their portrayal of the role. All conflict simulation cases that use role-playing need a forum in which the conflicting interests can debate. For the burger case it is the annual stockholders' meeting, for Chatham it is a meeting of the state water board. One can simulate many issues and bring the role play to culmination through mock congressional hearings, judicial hearings or even with mock appearances on a television talk show.

Conflict simulation cases are the best way to synthesize the reading/research, analysis, and writing/speaking elements of philosophical ethics, especially for students who tend to work best with concrete, problem-focused issues before them. They are pedagogically risky in that a few bad apples can truly spoil the experience for everyone, but when classroom sessions are well managed, there is no question that students learn far more from participating in one of these simulation exercises than from twice as much time spent in didactic learning. Yet I seldom use conflict simulations in my course now. The reason is that they are too expensive, not in cash money, but they take too much of the instructor's time to orchestrate and manage. Neither are they well appreciated by students. The work that an instructor invests in conflict simulation cases is unlikely to be repaid with good teaching evaluations. "The prof just stands in the back while we do all the work," is a typical comment. As I have been asked to teach progressively larger classes (100 students is now the norm) the prospects of using conflict simulation cases seems increasingly remote. I acknowledge the educational power of conflict simulation cases, but find that all the incentives and rewards for college teachers are stacked against them.

The third alternative is to use *decision cases* which describe a situation in which an individual must make a crucial, value laden decision. The reigning dogma is that decision cases must be real, historical examples, and that they should focus on the choice situation of a single individual. The case materials should include an accurate representation of information that the individual who faced the choice must have to work with, and should describe other parties only to the extent that they function as elements that bear on the key decision. This method of case study teaching has been very successful in business ethics and medical ethics situations. I have used a decision case on dairy expansion to salutary effect in my agricultural ethics class, but I see two limitations. First, I would note that both executives and physicians have an extraordinary degree of power and latitude to make choices when compared with the average person. Many of us (and farmers are like us) must express our values through attempting to accommodate our practice to situations that are far beyond our control. Second, decision cases are intrinsically ill-suited to a discussion of the ethical principles that should underlie or inform a public policy, or even a set of social norms that must evolve through informal cooperation. In themselves, decision cases are great, but the mythology of autonomous decision makers that goes along with them is heightened by the dogmatic insistence on realism and personal choice. They complement conflict simulation cases, but they should not be allowed to replace them.

Technology

Whatever the subject matter or the case methodology, how should agricultural ethics be delivered? The philosophers idea of ideal teaching technology is a log. Two people sit at opposite ends and talk. Not only is the teacher-to-pupil ratio of log technology prohibitively expensive, however, there is something to be said for giving students from diverse backgrounds an opportunity to talk with each other. The standard discussion class is a good technology for achieving this, but both efficiencies and even educational quality can be enhanced when electronic technology is used to broaden the audience (and the conversational pool) of students to include those from other campuses around the nation.

In 1992 I experimented with a satellite broadcast using the AG*SAT television network. A live studio broadcast of my agricultural ethics course, complete with students, was broadcast to about ten remote sites, three with live student interaction and three others with students accessing the broadcast on a tape delay basis. For those who were involved in the live interaction, the result was excellent. Students from Tennessee, Illinois, and Texas were engaged in debates with one another, and the differences of perspective from even these three relatively conservative states was impressive. The technology got in the way of strong interpersonal interaction with students in the studio classroom, but on other points it was a success. However, the AG*SAT broadcast had limits. It was feasible to link three off-site locations on a real time basis, but it is doubtful that the number could be expanded much beyond five or six. Instructors at each remote location were responsible for monitoring the class and for evaluating student performance. So four instructors taught relatively small groups of students (ranging from fifteen to fifty) in an educationally sensational environment, but with an additional $60,000 worth of investment in satellite hook-up and production time. There were also significant planning and organizational costs for the time of instructors who participated. In a nutshell, it's too expensive.

An alternative is casual dial-up audio-visual interface. Systems that operate over ISDN telephone lines were tested in the Fall of 1995. The technology turns an ordinary desktop PC into a video-phone that can connect any two points linked by the high-capacity ISDN telephone lines that are becoming the standard for data networks. Here, coordinating is as easy as making a pre-arranged telephone call, and costs are virtually negligible —comparable to a long distance call. Face to face communication can be augmented with videotape, with a second camera for a visual display, and with a typical PC paintbrush apparatus that allows one to mark up the video screen the way play by play sports announcers do on television. As tested in 1995, PCs with large monitors were installed in classrooms, then the guest lecturer "visits" from the comfort and convenience of his or her office.

This technology works. It requires no more advance planning than an ordinary telephone call, though of course if one is using it for a formal course some minimal advance planning is required in order to ensure that one's correspondent is "home" and not otherwise engaged. I used ISDN casual dial-up to teach class sessions at two campuses, one in Colorado and one in Wisconsin during 1995, and more than a dozen sessions were taught during the test with instructors based all over the United States. As more individuals and institutions have the equipment to dial up, it will become increasingly useful. The 1995 version of the technology did not support the high-tech graphics, audio and picture quality that one expects from television, and picture resolution did not provide a robust image of anyone sitting more than fifteen or twenty feet from the camera. The low cost and convenience of casual dial-up more than compensate. Will it substitute for hiring philosophers on campus to teach and develop agricultural ethics? Assuming that faculty members maintain current teaching commitments at their home institutions, one can envision them doing six to ten of these extended guest-lectures via dial-up each semester. If all ten or so of the philosophically trained agricultural ethicists participated, there would be up to a hundred single class sessions of visiting lecturers available each term. If that number is tripled to include the excellent (but not philosophically trained) people from agricultural colleges that could contribute to this effort, the total is expanded to three hundred sessions per term. But if only five percent of the agricultural classes on each campus in the United States and Canada decided to devote only one class period to ethics, the demand for these sessions would easily expand into the thousands.

In short, technology will clearly be changing higher education faster than anyone suspected only a few years ago, but decision leaders have probably not fully appreciated the true costs of these changes as yet. Agricultural ethics is in one respect a model opportunity for exploring remote instructional techniques in light of the short supply of expertise and the potentially large demand. However, as administrators come to grips with the high costs that are associated with experimentation in educational technology, the marginal status of agricultural ethics in most agricultural curricula may make this area appear less attractive. Even in the technologically most-optimistic cases, the field needs two to three times more committed specialists than it can now be said to have even under the most liberal interpretation of what it takes to be an expert on agricultural ethics.

6

Agricultural Ethics in Rural Education

Although previous chapters have stressed research, agricultural ethics is a philosophy emphasizing ethical issues associated with food and fiber production, processing, distribution, and consumption. Farming and ranching have historically been viewed as morally praiseworthy activities, and rural citizens have been promoted as models of morality. Now careers as farmers or as representatives of agribusiness firms inevitably involve practices that are currently and foreseeably embroiled in controversy. Norms and values are and will be a substantial component of such controversies. As such, familiarity with agricultural ethics may be a substantial asset for residents of rural areas who contemplate such careers. What is more, the adage, "If you eat you are involved in agriculture," is particularly apt where agricultural ethics are concerned. All citizens depend upon secure agricultural production, and all affect the practice of agriculture through their consumer decisions and through political action. There is, therefore, good reason for all students to be cognizant of issues in agricultural ethics.

Rural America has come to be perceived as a locus for ethically and politically contentious activities following the publication of books criticizing agriculture. Rachel Carson's *Silent Spring* was published in 1962. It was arguably the first in a series of books that raised questions about methods of agricultural production that were being adopted in the industrialized world. The succession of titles in this list includes E. F. Schumacher's *Small Is Beautiful* (1973), Frances Moore Lappe's *Diet for a Small Planet* (1971), Wendell Berry's *The Unsettling of America* (1977), Wes Jackson's *New Roots for Agriculture* (1980), Orville Schell's *Modern Meat* (1984), Michael W. Fox's *Agricide* (1986), and John Robbins' *Diet for a New America* (1987). Each of these books includes empirically questionable accounts of industrial agriculture, but each also presents an alternative view of agriculture's goals. These alternative views are imbued with moral qualities alleged by the authors to be superior to those of institutional, industrial agriculture. Often, these alternative views of agriculture are presented as attempts to recapture the moral qualities traditionally associated with agriculture and with rural communities. As such, although agricultural philosophy includes substantive discussion of issues ranging from water pollution to animal rights, each of these specific issues is often defined as part of a more general attempt to specify the ethical significance of agriculture and country life.

This chapter will examine the way that several authors have approached a philosophical account of farming, ranching, and rural American life. Contrasting philosophical opinions on the significance of agriculture can be organized as competing answers to the question, "Why is agriculture special, or why does production of food and fiber differ in ethically significant ways from production of industrial commodities or services?" One answer, of course, is that agriculture is not special, and that agricultural producers should be viewed exactly as we view owners, operators, and workers in convenience stores, manufacturing, or the dry cleaning industry. There is no reason to think that engaging in such lines of work imbues these people with morally superior character traits, nor do these industries seem to call for ethical considerations that distinguish them from the general pattern of human activity. Philosophies that present reasons for thinking that agriculture is special can be classified as "agrarian." This chapter will examine several varieties of agrarianism and will conclude by reviewing some topics in agricultural ethics that might be an appropriate focus for rural educators at both secondary and collegiate levels.

Traditional Agrarianism: Jefferson and Emerson

As already noted, the specific subject matter of agricultural ethics is quite broad and comprehensive. It includes environmental concerns related to agricultural chemicals, the sustainability of agriculture, the impact of technology and public policy on the size and economic structure of farms, the process and procedure for certifying food safety, the responsibility of wealthy nations and individuals to aid the poor of other countries (either through food aid or technical assistance), and the emerging relationships between population growth and food production. Blatz (1991) has published a comprehensive anthology on agricultural ethics which includes representative readings on all of these topics. Key issues related to agricultural technology are discussed in the essays collected in Thompson and Stout (1991); surveys of agricultural ethics issues as they relate to college curriculum are found in Thompson (1988c, 1990a, as well as the preceding chapter).

Although work now being produced in agricultural ethics may provide future generations with more-explicit and careful attention to normative controversy about agriculture, the current generation of adults may believe that rural residents, particularly farmers, are more likely to exhibit ethically praiseworthy conduct and more likely to base action and decision on ethical principles. In one manifestation, agrarian ideology has maintained the notion that farm families are more likely to be guided by principles of ethics than are others, and that because farming is morally significant, agriculture should be given special consideration in matters of public policy. Bonnen and Browne (1989) argue that an unwarranted belief in agrarian values was partially responsible for dysfunctional farm policies, and in a recent book, Browne, Skees, Swanson, Thompson, and Unnevehn (1992) have offered a comprehensive review of agrarian beliefs and their relevance to farm policy. The agrarian belief that agriculture is morally special is therefore both a topic for agricultural ethics and a more general way of thinking about the social role and function of agriculture that determines how scholars of rural life will frame the specific questions of sustainability, farm structure, and environmental quality. Since students in rural schools are the designated carriers for this peculiar deposit of virtue for the generation to come, it is particularly appropriate for rural educators to examine the themes of agrarian philosophy.

Without question, Thomas Jefferson is the fountainhead for American agrarianism, yet his ideas are often seen through the lens of a highly romanticized vision of agriculture (not to say America) that was not typical of his thought or of his time. Jefferson experimented extensively on his Monticello plantation, and his record of this activity reflects an attempt to increase the economic and biological productivity of his farming operations (Betts, 1953). In this regard, Jefferson is far more like the modern agricultural scientist than the contemporary agrarian critic of industrialized agriculture. To be sure, the production methods of Jefferson's time were a far cry from the energy and chemically intensive production methods of today. There is nothing in Jefferson's writings to suggest that he would endorse industrialized agriculture, but neither is there much to suggest that he would condemn it.

The writings that form the basis for "Jeffersonian agrarianism," the view that farmers are morally special, are surprisingly few and narrow. The most cited passages are from *Notes on the State of Virginia,* in which Jefferson describes farmers as "the chosen people of God" but goes on to qualify this description with the disclaimer, "if ever He had a chosen people" (1984b, p. 290), and from the 1783 letter to John Jay in which farmers are described as being particularly valuable as citizens (1984a). Contemporary interpreters of Jefferson have read these passages as endorsing the view that farmers would naturally develop moral virtues, where city dwellers would not. As such, it has been thought vitally important to maintain an extensive rural population as a conscience, so to speak, of the national political will.

Theodore Roosevelt, for one, described just such a vision of farming's moral purity, as did Eisenhower's influential Secretary of Agriculture Ezra Taft Benson, though neither attributed the view to Jefferson. More recently, Wendell Berry has placed his interpretation of Jefferson into a context which invites the reader to equate Jeffersonian agrarianism with a belief in the moral superiority of farmers, though detailed examination of Berry's ideas reveals a more subtle argument. Philosopher Jean Jacques Rousseau gave impetus to the idea that people who derive their living from nature more perfectly realize the implicit human capacity for virtue. His first development of the idea of the noble savage, the *Discourse on Inequality* (1755/1984), was published in 1755, but Rousseau's mature ideas on education and moral development were more closely contemporaneous with Jefferson's own writings. Jefferson himself may have believed that farmers were more virtuous, but both Jefferson and Rousseau stress the role of conflicting interests in their praise of farming. The teleological vision of a human nature fitted to the natural world, in conflict with society, was to come later.

Jefferson's remarks on the moral virtues of farming must be read in the context of constitutional debate. Jefferson was an advocate of democracy, a cause which was opposed by a spectrum of ideologies and interests. A traditional argument against democracy stressed that common citizens would be tempted to waste public resources on unsustainable policies. The fear that people will demand benefits while refusing the taxes needed to supply them is familiar to our own time. Philosophers such as Plato and Hobbes proposed this criticism of democracy as part of an argument supporting the need for absolute monarchy, but political thought in eighteenth-century Europe modified the criticism by suggesting that any landowning class would have a personal interest in maintaining responsible public policies. Of course, in Europe the landowning classes were titled noblemen, so the argument supported a constitution in which power is vested in a body such as the English House of Lords. Opponents of democracy were advocating a similar arrangement for the emerging United States.

Jefferson's passages on agriculture point out that in the United States, the landowning class is not noblemen, but smallholding farmers. Though many states, notably Jefferson's own Virginia, contained large plantation farming systems, many freeholding small farmers were interspersed among these plantations throughout the original thirteen colonies. Northern colonies were dominated by small farms. If the landed interests of dukes and earls could be relied upon to check their irresponsible impulses, why couldn't the same be said of small farmers? Jefferson's letter to Jay turns an argument against democracy upon its head, introducing considerations which make the same philosophical assumptions support a broad distribution of political power. This strand of argument is particularly evident in passages where Jefferson stresses that it is as citizens that farmers are "particularly valuable" (letter to Jay, see Jefferson, 1984a).

Although agriculture is thought to be special and was clearly an activity that Jefferson relished personally, in the crucial philosophical passages it is the high percentage of the population deriving their livelihood from land holdings that produces the value of agriculture for a constitutional democracy. It is worth repeating and stressing three points about Jeffersonian agrarianism:

1. Jefferson says relatively little that would support the view that farmers have a stronger moral character than non-farmers. Indeed, the main philosophical arguments assume that farmers will act in pursuit of their own best interests.

2. Farmers' personal interests coincided with the duties of citizenship because farmers needed military protection and a strong economy to assure their livelihood. They were not thought to possess either financial assets or personal skills that allowed mobility. They were tied to a specific geographic location and were thereby committed to the long-term future of the state in whose boundaries their farms fell.

3. Jefferson thought that American democracy would be particularly vulnerable to special interests during its formative stages. He saw reliance on the farming population as short-term protection from special interests, as protection that was sufficient to overcome the objections raised by the opponents of democracy. He knew that more broadly based social institutions would be required to make democracy work as the United States became more urban and industrial.

The striking implication of the second and third points is that while they provide excellent philosophical justification for the constitutional points at issue between Jefferson and the opponents of democracy, they presuppose an agricultural economy substantially different from today's. Farming is now both capital- and skill-intensive, and with less than 2 percent of Americans employed on farms, our society has clearly long passed the point where social institutions must be relied upon to protect democracy from capture by special interests.

Although Jefferson is the most frequently mentioned agrarian from the pantheon of America's past, a different intellectual tradition supplies the substance for contemporary agrarian views. European emigrants who came to North America to escape religious persecution may have understood their experience as a purifying "errand into the wilderness," and they clearly viewed the European societies they fled as corrupt and immoral. In coming to the North American continent, they constructed their experimental communities on moral principles derived from their religious beliefs. They were obliged to be farmers by economic necessity and made a virtue of this necessity by developing folkways that celebrated and integrated farming with their religious quest (Miller 1956; McDermott 1987). Nevertheless, the religious beliefs of American colonists were not, for the most part, characterized by the view that farming (or otherwise deriving one's livelihood from nature) would produce a more virtuous individual or a more just society. The view that nature itself could serve as template and stimulus for virtue came later. Indeed, to the extent that immigrants to the New World were religious, they saw their task as one of civilizing a wild nature in terms of the religious creeds they brought with them. By the nineteenth century, however, the principal elements of American naturalism had emerged. Ralph Waldo Emerson is both an originator and a typical figure of the views that undergird a romantic vision of agriculture.

Emerson's agrarianism has been concisely exposited in a recent paper by Robert Corrington (1990). According to Corrington, the underlying philosophical vision is one which first defines morality as a fulfillment of human potential, then couples that precept with a naturalistic interpretation of human potential. The first tenet is one that has a long and distinguished philosophical pedigree. Aristotle's *Nichomachean Ethics* (1962 ed.) describes the good life as one that most completely succeeds in developing and maintaining the virtues. Virtues are understood as ends to which humans aspire throughout their lives. Furthermore, attainment of virtue is sought not because virtues are character traits that serve as instruments for obtaining or producing some external good, but because attainment of virtue is thought to be the principle which integrates a person's life activity, giving meaning and order to life in the most fundamental way.

Like Emerson, Aristotle had a strongly naturalistic interpretation of the ends to which humans should aspire. Virtue was thought to be implicit and native to human potential. By Emerson's time, however, Aristotle's teleological account of ethics had been supplemented by neoplatonists and Christian thinkers who stressed idealistic or theological accounts of human purpose. For Emerson, the primary philosophical task was to reconstruct a naturalistic philosophy of human purpose. Among Emerson's immediate precursors, those who had adopted naturalist views included Adam Smith, David Hume, and Jeremy Bentham. This group of philosophers defined ethics as an attempt to satisfy ends defined as the satisfaction of biological and psychological drives. In their theories, virtues were understood simply as habits that tended to maximize the satisfaction of such needs, and the possibility of one person failing to experience the pleasure or pain of another was thought to pose a problem for motivating ethical action. The British naturalists rejected the view that ethics was an attempt to realize potential, characterizing theological accounts of potential as socially dysfunctional ideologies designed primarily to protect clerics and aristocrats from the forces of social change. By contrast, Aristotelian virtues are not instruments for achieving satisfaction, but the very substance of human striving, albeit in unrealized form. As such, many of Emerson's essays discuss how a more sophisticated concept of nature was needed to rehabilitate teleology.

Corrington discusses how Emerson's new vision took the form of eulogizing human capacity for creative expression, a capacity most fully realized by the poet. Emerson supplanted the idealist aesthetic of pure form with open-ended pursuit and realization of the naturally latent and explicit. Contrary to those who took aesthetic perfection to require idealization aimed at eternal and unchangeable forms, Emerson saw poetic creation in terms of an inner voice, an inner vision. This inner voice was experienced as personal and contingent but was, in fact, universal and when expressed authentically, capable of evoking meanings true for everyone. The true poet attends to this voice from the self. Life, however, is a process of coming to hear and finally to speak from an authentic self, and this process will be impoverished and perverted by attempts to imitate others' authentic expressions. As such, the city, the salon, and the university do not provide an appropriate milieu for the aspiring poet, understood as a metaphor for all humanity. Nature itself is the milieu in which natural capacities of human expression will be most fully realized, and the farmer is a far better exemplar of the moral quest than is the false poet, the city intellectual, or the imitative artist.

Emerson's essays, as well as those of Thoreau and other nineteenth-century transcendentalists, are filled with praise for those who live in the presence of nature. No one is more emblematic of this life than is the farmer (Emerson, 1904). Emerson's praise for farming is imbedded in an ethic of personal fulfillment. This ethic is itself imbedded in an aesthetically authentic expression, for which authenticity is understood as unforced, non-imitative activity in pursuit of self-directed goals. The yeoman farmer lives in the full presence of nature. The farm family organizes all life activities around nature's cycles and nature's demands. As an entrepreneur, the farmer works for self, rather than for wages, and the farm is an authentic creation worthy of praise and admiration. Emerson's praise of farming presupposes ethics and aesthetics that reject both anti-naturalism and the biological/psychological determination of human needs posited by Hume, Smith, and Bentham. Emersonian agrarianism emerges as a philosophical thesis distinct not only from other nineteenth-century approaches to ethics, but also from Jefferson's more politically pragmatic praise of farming.

Wendell Berry's Agrarianism

Wendell Berry is a prominent contemporary critic of agriculture and more important, a critic of culture who offers an indirect and subtle argument for valuing farms and farming. The basis for his beliefs about agricultural values and issues is continuous with and indistinguishable from his broader philosophy of place, time, community, and values. Berry's philosophy has several elements which are consistent with the American philosophical tradition: his naturalism, his celebration of the ordinary as a spiritual source, his emphasis on the importance of making connections, and his respect for nature as both sacred and useful for living. Yet contrary to the American tradition that celebrates the redemptive possibility of an open future, Berry advocates the superiority of the circular or cyclical model of time. American philosophy and culture have been dominated by a linear model of time evident in the journey motif common to American literature, the westward expansion of early European settlers, and the frequently recurring ideal of progress. His proposition to improve and in a sense to save American culture and agriculture is to abandon linear time in favor of circular time or recurrence. Recurrence is a foundation for further arguments for the importance of farming on a small scale, the value of human and animal labor on the farm, and the need for preserving the fertility of the land despite any short-term economic benefits that could be gained from its abuse. The feature of recurrence most relevant to agriculture is the preeminence of place—being aware of the native and local and the value of remaining in the place and at home. The small-scale family farmer, understood as cohabitant and steward of the land, serves as a model of human virtue.

It is evident in Berry's writing that he believes human beings are fundamentally natural creatures, meaning that people and their social forms are thoroughly intermingled in nature and nature intermingled in people and society. Furthermore, Berry stresses how all of nature is connected in a closed organic system. He states:

> Obvious distinctions can be made between body and soul, one body and other bodies, body and world, etc. But these things that appear to be distinct are nevertheless caught up in a network of mutual dependence and influence that is the substantiation of their unity. Body, soul (or mind or spirit), community, and world are all susceptible to each other's influence, and they are all conductors of each other's influence. (1977, p. 110)

Berry should not be called a naturalist in the strictest philosophical sense perhaps, because he does not deny the existence of non-natural entities. Overtones of traditional western religion do often appear in his poetry and essays, which might even indicate some belief in the supernatural. However, his references to Creation and God never imply that life is a prelude to greater posthumous existence but are affirmations that the natural world is also spiritual. Berry leaves the question of the supernatural and non-natural existences unanswered, perhaps because the answer is unimportant.

What is certainly important is the realization of the material state of man, with his reliance on nature for the necessary resources for living, as well as the potential of nature to be a source of spirituality. It is clear that interaction between humans and nature is both physically necessary and potentially consummatory. His poetry relates the beauty, practicality, and sanctity of human interaction with nature and also the spirituality inherent in nature. In a poem in *Sabbaths* (1987) he affirms the spiritual presence of trees:

> *Slowly, slowly, they return*
> *To the small woodland let alone:*
> *Great trees, outspreading and upright,*
> *Apostles of the living light. (p. 95)*

Berry's essays also establish the connection between human work and spirituality. He indicates that the most spiritually fulfilling actions are practical actions. Working, in the sense of performing routine household tasks important for living in reasonable health and cleanliness, is not a burden or a necessary evil. Rather, the routine task is a possible source of spirituality.

The importance of labor to spiritual health brings out an important component of Berry's philosophy, the predominance of role. Morality is centered in the kinds of roles people fill and not in individual actions. Although Berry is quick to point out moral failings in American society, these are not moral failings in the Kantian sense of people having failed to perform their reason-sanctioned duty. The kind of moral failure he finds in America is a failure of character. Americans are too greedy, too wasteful, too lazy, too future-oriented, and too ostentatious. They are not thoughtful enough about what they are doing but are busy rushing through daily activities. What Berry calls the crisis of agriculture, he diagnoses as a fault in the contemporary American character.

Berry believes that forming the proper relationships of nurturing and caring for farmland, wildlife, and one's own neighborhood are essential parts of the health of the community as well as the health of its constituents. Health is a recurrent theme in Berry's writings. Berry at one point equates health with independence (1977, p.183), but a closer reading indicates that he means health is independence from what is foreign. A healthy farm should be independent of manufactured fertilizers, for instance, but should not exist independent of local weather conditions. In other words, health is maintenance of the proper connections. Proper connections in this regard can be the connection between a farmer and his crops or livestock, between a man and wife, or a factory worker and his product. Being connected with the earth means one is committed to protecting the land from undue damage from wind or water drainage as well as a commitment to taking food and shade from the land in order to accommodate human life. It is an organic relationship.

All these themes—nature, spirituality, work and health—are framed by temporal recurrence. In "Discipline and Hope" (1972), he reveals that thinking of time as circular is obvious for him and that the situations that man puts himself into are eternal. The same problems that arise today were problems years ago, and the solutions to them are meliorations, not permanent solutions. Linear time—the time of moments arrayed infinitely into the future—he criticizes as destructive in two ways. First, it is morally dangerous because it leads people to thinking that sinfulness today can be rectified in the future. He envisions the thinking of an entrepreneur as typical of American linear temporality. The entrepreneur believes some local destruction (of land or minerals or air quality, for example) must occur in order for there to be progress made and money earned. The entrepreneur's thinking is that this destruction is justified from the perspective of the future, considering the monetary gains and growth of the economy that counterbalance the environmental or social harm. This is understandable with the linear notion of time, Berry believes, because the present has no value, or no inherent value, except that it possesses the capabilities and resources to create the future. Second, linear time robs action of its meaning. Berry believes that adherents to a concept of linear temporality think of actions as means to future ends. The question "What is this good for?" arises about any proposed action, and the question can only be answered by reference to some good not yet at hand. This is necessarily a rejection or dismissal of the inherent worth of possessions or actions. By thinking of something as worth something else, its value is depreciated, and there is nothing of sacred worth.

Berry rejects the notion of linear time because it is incomplete. Circular time on the other hand is complete. The deficiency in linear time is due to its ignorance of other values than those that are ultimately justifiable in the future, values other than eschatological values. The values that Berry believes are slighted by the linear time vision are the ones he finds most valuable, the value of doing good work, respect for nature, maintaining and promoting health, respect for death, and most importantly, respect for the continuance of life. The circular vision of time encompasses these values because the circular vision of time is by Berry's definition a biological and organic vision of time. The recurrences that it postulates are those of the cycles of birth and death. One consequence of circular time for agriculture is that great value is placed on permanence and remaining in place. Clearly, if time is not linear and life is not a journey, there is no place one needs to go. Instead, one should educate oneself about the local land and community so as to be better able to adapt to the environment and treat it respectfully. Knowledge of the local geography, geology, wildlife, and human community becomes essential for connection to the land and one's possible success in living. This emphasis in Berry's philosophy is not a license for moral egoism or abandonment of concern for other cultures and practices, nor does it disregard history. On the contrary, Berry's paradigm of circular time encourages the idea that the conditions under which man lives are roughly the same at any time or place with certain local adjustments for geography and history. Understanding of other cultures' practices is helpful for improving the practices of foreigners and natives. In fact, what is important about respecting the local and present is that it is a specific and tangible tie to the earth. What

Berry sees as problematic is an American culture which has no ties to the land. The danger of a nomadic society where people are tied mainly by the corporation in which they work and not by local connections is that they will not have anything at stake in the place where they live. Where they live will not be important because for all they care it could be anywhere. From this it follows that people will tend to misuse the land because there is no moral motivation for them to care for it. Long-term financial pressures, like the depreciation of the value of farm land with the decline of its fertility, are insufficient to motivate the proper attitudes of tenancy because they are merely financial. Proper care of land, he believes, can only be secured when it is believed morally correct, natural, and beneficial.

The import of Berry's cultural criticism on the teaching of agricultural ethics is that agricultural problems and solutions should be seen as cultural problems. What are often regarded as specifically rural concerns, such as large-scale migration of rural inhabitants away from the farms into cities and suburbs and the consequent dissolution of rural community structures, are in fact concerns for all of society. Agricultural issues as well as ethical issues are found to be based on issues of character. Several possibilities exist for addressing questions in agricultural ethics and rural life consistent with Berry's analysis of agriculture. One possibility is the investigation of the differences between nomadic and sedentary societies regarding their treatment of land. The social consequences of agricultural innovations, governmental regulations, and traditional farm practices could be examined, with an emphasis on what values or character traits they support. Study of agricultural economics could include discussion of externalities and the implicit goals of economics as opposed to the goals of agriculture. In all, the cultural aspects of agriculture should serve as a foundation for teaching agricultural issues and as a justification for their importance.

Agrarianism and Rural Education

In the preceding sections, some of the key themes of agrarian philosophies have been identified. Each of these philosophies specifies a special role and function for agriculture and rural life. In Jeffersonian agrarianism, it is the way that a broad-based pattern of landholdings moderates political tendencies that are inimical to democratic government. In Emersonian agrarianism, it is the way that life in the presence of nature is conducive to the creation of authentic expressions of human spirit. In Wendell Berry's agrarianism, these two themes are integrated in light of threats to personal and social health that derive as much from cultural as from environmental catastrophe. Each of these visions provides a comprehensive framework for the self-understanding of rural culture. Furthermore, they define rural culture as distinct from urban culture, and thus present a framework from which to evaluate the goals of rural education as uniquely determined by the character of the rural environment.

At the same time, it would be irresponsible to conclude a chapter on the role of agricultural ethics within rural education without noting how a far less comprehensive and totalizing philosophical vision of agriculture and rural life might also be appropriated within curricula for rural education. Philosophical materials on the role and function of rural life are often complex and presuppose vocabulary and intellectual development for which high school students are unprepared. Educators focusing on K-12 have several options available for incorporating agricultural ethics into their courses. One is to incorporate agricultural ethics into course materials focusing on history or literature. This approach is examined below in application to the novel *The Grapes of Wrath* (Steinbeck, 1939/1979). A second option is to take up controversial issues in agriculture within a science or social studies curriculum. This second option is reviewed in considerably less detail in the closing paragraphs of this chapter.

Literary and Historical Approaches to Agricultural Ethics

Agrarian and ethical themes are already represented in literary and historical studies to some extent. Teachers who are sensitive to the varieties of agrarian philosophy can certainly emphasize them within the framework of existing materials. John Steinbeck's *The Grapes of Wrath* is a standard text for many high school students, and it can be adapted to stress philosophical themes in addition to those traditionally taken up in English classes. Steinbeck wrote his novel during a crisis period of American economic history. Along with Archibald Macleish and Dorothea Lange, Steinbeck took the disenchantment with capitalist values that was widespread among 1930s literary intellectuals beyond familiar urban settings and created a narrative that probes the internal contradictions of American rural life. At its publication, the book was widely perceived as a political and moral critique of the economic forces that were transforming American agriculture. The Joad family's fictional chronicle is typical of economic events that have affected American farm families throughout the twentieth century (not only in the Dust Bowl years), and Steinbeck infuses their story with social and political observations that raise fundamental questions about private property, public policy, and class consciousness. As such, any reading of the novel that minimizes its political commitments neglects one of its most impressive accomplishments.

In *The Grapes of Wrath,* Steinbeck presents a philosophy of human nature through the character studies of the Joads, their neighbors and other people they encounter. The political and economic themes of the novel are presented partly in the situations of the plot and partly in the chapters that establish context. Chapter 5, for example, begins with an anonymous encounter between owner and tenant that outlines the economic imperatives that have caused the ouster of Dust Bowl families. The chapter continues by describing the tractor driver as one who "could not see the land as it was," and who "loved the land no more than the bank loved the land" (p. 37). Here, the new mechanized farm technology separates the tractor driver both physically and cognitively from an economy that sees the land in terms of place, of home, of stewardship. An economy that sees the land merely as a factor of production takes its place. Throughout this and other such chapters, Steinbeck explores the relationship between people of a given social role (owner, tenants, employers, and businessmen) and the broader social, economic, and ecological environment. In episodes with used car dealers, cafe owners, and California farmers, the incentives of a profit-seeking economy bring people into conflict with traditional ways of life. In the brief but poignant Chapter 14, the root causes of the Dust Bowl itself are tied to the quest for efficient production. The tractors, brought onto the Great Plains to increase commodity production, have broken the land and cast its occupants out on the road. Steinbeck concludes the chapter with this passage:

> If you who own the things people must have could understand this, you might preserve yourself. If you could separate causes from results, if you could know that Paine, Marx, Jefferson, Lenin, were results, not causes, you might survive. But that you cannot know. For the quality of owning freezes you forever into "I," and cuts you off forever from the "we." (1939/1979, p. 166)

As students come to a complete understanding of such passages, they will naturally be led to an encounter with crucial issues in agricultural ethics.

Since the turn of the century, rural areas dominated by family farms had been undergoing a transformation all across the United States. Steinbeck correctly perceived that all of American agriculture was beginning to look like the large-scale monoculture of California that he knew well. Popular opinion blamed individual farm failures on bad luck and bad management. Firms in every industry were increasing in size; agriculture appeared to be only another instance of "bigger is better." The Dust Bowl was only the most dramatic instance of this transition, and it too was widely thought to be a natural disaster, not a sign of fundamental problems in the nation's agricultural economy. Steinbeck rejected this explanation, and his portrayal of the Joads' predicament always cites poor production choices and bloated markets as the reason for their failure. References to natural disaster or divine will are relegated to the Okies' self-perception of their plight, never to the narrator's point of view. Steinbeck did not, however, present more than a vague account of how the Dust Bowl could be understood as the result of economic, rather than natural, forces. Neither did he show why family farm failures were endemic to the structure of American agriculture rather than consequences of the financial stress brought on by the Dust Bowl and the depression.

The philosophy at work in *The Grapes of Wrath* can be elucidated by study of three main points that are either included in the novel itself or that are the object of occasional allusions. First, the Joads' eviction from their Oklahoma home must be understood as the result of socioeconomic forces (including technological innovation) rather than bad fortune, the unlucky consequence of a natural disaster. Second, the Joads' experience as migrant laborers in California must be seen as a sociopolitical class transition, in which the entire population of Okies must cease to think of themselves as yeoman farmers and must take on the self-conception of the unskilled wage worker. Finally, the ambiguity of Steinbeck's reaction to these events must be recognized, and his reaction can be used as a challenge to contemporary thinking about social justice. Though political philosophy is implicit in passages throughout the novel, stressing these three points sharpens the focus and allows a more deeply grounded philosophical reading of the political theme.

The "structure" of agriculture (e.g., the relative size distribution of farms, patterns of ownership and tenancy, relative percentage of off-farm income, etc.) has undergone a dramatic transition since 1900. At the turn of the century, approximately 70 percent of Americans lived on farms, while today the number stands at 3 percent. While there are many factors that have influenced this transition, the unwanted loss of a farm home through bankruptcy and eventual eviction has not been atypical, and many families have voluntarily abandoned farming because they saw in advance that such a fate was in store. Although *The Grapes of Wrath* begins after many of the key economic events have transpired for the Joads, two concepts from economic theory help explain what might have happened to them in a way that is fully consistent with Steinbeck's text.

The first concept is that of the technological treadmill: Agricultural technology increases farm productivity, but this in turn lowers prices, forcing individual farmers to run faster just to stay in place. Much of American agriculture was mechanized for the first time in the years from 1910 to 1925. The introduction of tractors and early automated harvesters increased productivity on American farms. This increased productivity was a boon to farmers in the years that U.S. farm exports went to feed a war-torn Europe. The new equipment, however, was more efficient for relatively larger farms. Eventually, farmers who had accumulated debt to invest in new equipment were caught in a spiral of falling prices as mechanization became widespread and productive capacity exceeded demand. These farms either failed or were bought out by larger farms which could expand total sales to keep up with falling unit prices.[1] This seems to be what has befallen the Joads, who once owned their land, mortgaged it, lost it to the bank, became tenants, and are being evicted, as the novel opens, to allow for farm consolidation.

The second concept is that of the prisoner's dilemma, borrowed from game theory by resource economists and rechristened "The Tragedy of the Commons," by Garrett Hardin (1968). The idea here is that each individual user of a commonly held fragile resource has no incentive to conserve its use; indeed, conservation practices turn out to be doubly costly, since one sacrifices not only the short-run benefits of exploitation but also the long-run savings (actions of others deplete the resource anyway). The prisoner's dilemma accounts for a pattern of resource use in agriculture that, when combined with the effects of the treadmill, creates a one-two knockout punch for farmers of a particular region. In the Dust Bowl years, wind-borne erosion ruined farms without regard to a particular individual's use of conservation practices, and in more recent years, pumping of aquifers has ruined access to underground water for extravagant and conservative users alike. When the treadmill effect is added, the result is devastating.

When the events of *The Grapes of Wrath* are interpreted in light of these two economic concepts, it becomes clear that the Joad family is the victim of property rules and public policies that create or reinforce incentives for technological innovation and resource depletion. The

---

1. The treadmill concept is usually traced to the work of economist Willard Cochrane, who introduced it in a series of papers in the 1950s. See Cochrane (1979) for a non-technical discussion, and Cochrane (1985) for a discussion of its continuing influence.

treadmill and the prisoner's dilemma have been introduced into theory recently, so it is unlikely that Steinbeck could have had them in mind in 1939. Nevertheless, several passages allude to the socioeconomic forces in play, while bad luck or divine retribution are mentioned only as ideas the Okies use to explain their plight to themselves. Steinbeck seems to be favoring a political explanation of these events, though it is crucial to the class character of the Joads and their companions that they would not think in such terms (at least not in the early chapters). If so, a reading of the book that stresses each character's response to mere adversity (or even existential strife) fails to note that the political criticism (which reaches its peak in the California orchards) is a persistent and central theme throughout.

Even as farm tenants, the Joads retain a self-concept as yeoman farmers, individuals who produce the means of their own sustenance under conditions of economic independence and personal autonomy. The agrarian ethic of the yeoman farmer entails a resolution of the tensions between self-reliance and community, between stewardship and exploitation of the land. This theme in the novel presents a good opportunity to introduce agrarian views associated with Jefferson, Emerson or Wendell Berry. After they leave the farm, however, the Joads must sell their labor on the flooded California migrant labor market. They become proletarians, not autonomous Jeffersonian citizens. Despite the agrarian nature of their employment, migrant laborers may represent the purest form of proletarian labor to have existed in twentieth-century America. With only their labor to sell, the cherished norms of the yeoman farmer break down. Labor ceases to function as a guarantee of sustenance. Work cannot be done at the discretion of the worker, for wage labor is often an inadequate and unreliable source of income. Wages are available only under conditions controlled by the Farmers Association of the novel. Under these conditions, the Joads lose not only their individual self-respect, but also their sense of community as they are forced to compete with one another for scarce jobs.

Read in light of these observations, the second half of the novel is a study in the conflict between two types of class consciousness, two types of ethic. The ethic of the yeoman farmer requires sharing and camaraderie, but it also presupposes a status of personal dignity and self-reliance. As the families move West, the norms of sharing and generosity begin to erode as it becomes clear that each family will be in an increasingly weaker position to repay debts to others as time goes by. At the same time, more fortunate minor characters are drawn into an ethic that stresses protection of property rights and denies duties to bring aid to the poor. In short, we see the ethic that underlies rural

America being transformed from one in which basically independent equals spontaneously recognize a set of communitarian duties, to one in which property owners confront a larger class of wage workers. The workers, in turn, are placed in a competitive situation that precludes the formation of any but the most minimal social norms.

In spite of the clarity with which Steinbeck depicts the class transformation that overtakes the Joad family, he seems quite reluctant to accept a traditionally Marxist evaluation of their situation. Indeed, the main characters can never fully accept an ethic of class conflict. Tom Joad, for example, commits his second killing more from conventional loyalty to Casy than from any perceived revolutionary consciousness. The ambivalence with which Steinbeck views radical politics is even more apparent in the novel *In Dubious Battle* (1936/1979), in which agrarian agitators appear first as heroic figures but end up as men who lack a crucial capacity for human feeling.

Though hardly conservative then, Steinbeck seems unable to give up entirely on the work ethic of the yeoman farmer. Steinbeck's socialism and his critique of the power of private property are rooted in a philosophy that seems committed to the ideal of role morality. The notion that morality could be rooted in the work experience of agrarians occupied an important place in the thought of Jefferson and Emerson. The farm family is blessed because each member occupies social roles that make the interdependence of self, family, and public interest transparent. This is a philosophy that stresses the particular rather than the universal, as the individual is linked to the moral community by an extensive network of ties to specific others, family members, neighbors, and store owners in the nearby town. These relations cannot be generalized because they depend not only upon the specific people who occupy these roles, but the placement of the roles around the immovable geographic place of the farm, the land itself.

Agrarianism is an ethic of place, and the Joad family is out of place in California. When the family becomes uprooted, their inherited moral framework is subjected to stresses that it is not prepared to withstand. Steinbeck might have used this setting as a platform from which to characterize the agrarian ethic as self-deceptive ideology, or at least to promote an ethic of class conflict as an alternative. A true Marxist certainly would have seized this opportunity. *The Grapes of Wrath* always returns to the agrarian framework, however, for its most authentic and poignant portrayals and indeed closes not on a note of class consciousness but on a note of uniquely particularized and specific need.

This characteristic of Steinbeck's writing might lead one to see his underlying philosophy as nostalgic or sentimental, and that, perhaps, might be the appropriate assessment in the end. In emphasizing an ethic based upon particular persons sharing a common place, however, Steinbeck challenges the conclusions that a socialist ethic would find essential. He also eschews philosophical approaches to ethics that stress universal rules and impersonal assessments. Before rejecting Steinbeck's ethic as sentimental, it is important to assess the force of its opposition to philosophical systems that ignore, perhaps even repress, the way that specificity and particularity give rise to loyalties, to duties, and to a vision of the good.

Science and Social Studies Approaches

A somewhat less grandiose and complex discussion of agricultural ethics can be introduced using materials that take up controversial issues in agriculture. Developments in agricultural science and technology pose ethical challenges for people in agriculture. Such developments are the target of critics such as Rachel Carson or John Robbins. To date, pesticides and other agricultural chemicals have been the primary targets of critics, but mechanization of farm production, food safety, and the conservation of soil and water resources have also been noted. Even a representative survey of these topics exceeds the scope of the present inquiry. Interested readers are encouraged to consult the sources cited at the beginning of this chapter. The ethical implications of agricultural science and technology can and arguably should be taught as part of the science education component in rural education. Such coverage of ethical issues does not depend upon an understanding of the agrarian views sketched above.

Good case studies exist for discussing the ethical implications of science and technology. Public television's NOVA series has produced high-quality documentaries on subjects ranging from world hunger to the conservation of genetic diversity. A good overview of agricultural technology can be found in the film *Down on the Farm*. For the future, biotechnology promises many new developments for agriculture, but some of these will be (and have been) ethically controversial. Ice-nucleating bacteria and recombinant bovine somatotropin present two case studies of controversy in agricultural biotechnology that may be readily adaptable to high school use.

Ice-nucleating bacteria affect the formation of frost. Scientists have discovered that bacteria with a gene deleted from their genetic code have a reduced affect on the formation of ice crystals. They think that if enough of these gene-deleted (or "ice-minus") bacteria are distributed on a crop such as strawberries or potatoes, they will effectively lower the temperature at which ice crystals form on the top by as much as two degrees Fahrenheit. The potential economic benefit of protecting crops from freezing temperatures is great. Agricultural scientists have been pursuing the idea of using biotechnology to manufacture ice-minus bacteria by the millions, then distributing them on crops when temperatures are predicted to drop into the range where reduced ice nucleation could be expected to produce economic benefits.

Every laboratory test has been promising, but attempts to initiate field tests have met with unanticipated resistance from ecologists, environmental groups, and the general public. At issue is whether even limited field testing of ice-minus bacteria poses environmental risks. More broadly, critics want to see adequate procedures for assessing and mitigating environmental impacts from agricultural biotechnology. Scientists hoping to test ice-minus bacteria under field conditions faced a series of lawsuits and procedural challenges during the 1980s. These events delayed field testing for at least five years, and the negative publicity associated with ice-minus discouraged biotechnology companies from seeking early development and commercial application of ice-nucleating bacteria.

Bovine somatotropin, or BST, is a growth hormone that can now be produced in large quantities using biotechnology. Dairy scientists have long known that producers who administer BST to dairy cows on a regular basis can increase the production of milk per animal. Economic studies have indicated that if available at a cost effective price, BST would be widely adopted by dairy producers. Economists have also predicted, however, that commercial introduction of BST will favor large-scale dairy producers. Some economists have speculated that traditional dairy states such as Wisconsin may lose production to southern states where land for large dairy herds is more readily available. As such, critics who are interested in protecting small-scale family farms have taken up the cause against BST. They have been joined by others who are concerned about the environmental impact of large-scale dairying and by still others who have questions about the impact of BST on animal health and well-being. More recently, a few critics have raised questions about whether people should drink milk from cows that have been treated with BST. Although the scientific basis for this last criticism is virtually non-existent, critics who have raised it have been successful in attracting attention from consumer groups and large grocery chains. It is far from clear that BST milk will be well-received by food consumers, regardless of scientific evidence regarding its safety.

The case of "ice-minus," is particularly attractive for high school science teachers, since the National Association of Biology Teachers has prepared case materials targeted for the high school level. These materials illustrate the key issues in the case and sketch out roles that can be taken for a role playing exercise or debate to teach the key issues. BST has also been discussed extensively, with many of the key points collected in a book edited by Milton Hallberg entitled *BST and Emerging Issues* (1992). While some of the science and economics discussed in the Hallberg volume is presented at a college (or even graduate) level, many of the chapters will be accessible to high school readers, particularly in areas where dairy production is important. The same two issues, as well as others, are also discussed in two publications—a basebook and leaflet series—from Purdue University which take up biotechnology in agriculture. The basebook is appropriate for high school use, while the leaflet series presents very concise discussions of issues that may well be used even by students in lower grades.

Conclusion

Agricultural ethics is an emerging area for philosophical inquiry, but it touches upon themes that are familiarly addressed in American literature and history. Students who will be involved in or affected by agriculture will need a sophisticated understanding of how norms, traditions, and perceptions are applied in developing the science and technology for current and future food production. Consumers also need a better sense of where food comes from and how to evaluate food choices in terms of safety, nutrition, social justice, environmental impact, and even cultural tradition. These are topics that are seldom addressed in systematic fashion in current curricula. In light of their particular relevance to students in rural areas where agriculture is a leading industry, rural educators should be devising strategies for incorporating them into the curriculum at the earliest practical opportunity.

Part Three

Public Policy

7

Analyzing Public Policy: The Case of Food Labels

This chapter outlines a framework for policy analysis, and demonstrates how ethics bear upon each element of the framework. Contested issues in food biotechnology policy are used to illustrate the applicability of the framework for interpreting policy conflict. Although this approach addresses several of the key points where ethical concerns bear upon food biotechnology, the chapter makes no attempt to survey the full range of ethical concern. What is more, this chapter does not present a normative argument favoring one policy option rather than another. The idea that ethics requires a particular set of policies for food biotechnology is *not* argued here. Instead, the purpose is to examine how ethical arguments establish a burden of proof for policy evaluation. The thesis is that effective policy making requires an ability to understand how different types of ethical criteria bear on policy. Insensitivity to contrasting ethical approaches will only prolong policy conflict.

The framework is then brought to bear on the question, "What ethical considerations are relevant to labeling policies for food products?" This question does *not* equate labels with warnings. There are many ways to configure a labeling policy that do not imply health claims. The short answer to the question is that there are two kinds of ethical considerations. The first has to do with the use of labels as tools to produce ethically desirable ends such as good health and consumer satisfaction. The second has to do with the role of labels in protecting the principles of consent. While these two ways of evaluating labeling policies may converge, they may also indicate contradictory directions. The long answer to the question uses a general approach to ethics and policy to show why this is the case.

A Framework for Policy Analysis

Schmid (1987) presents a theoretical framework for policy analysis in which the laws, procedures and administrative decisions that serve as the instruments of policy are analyzed in terms of the incentives they create for key actors. Schmid's framework develops a public choice/transaction cost approach to public policy that permits an analysis of how informal norms and standard operating procedural interact with the formal apparatus of law and administrative decision making. Conventional economic policy analysis assumes that costs and benefits of policy can be computed simply by examining the impact of laws and administrative decisions upon production costs and consumer demand for regulated goods. For purposes here, the key insight of Schmid's approach is that he takes the formal apparatus of policy to be one component in an ensemble of laws, norms and standard operating procedures. The totality of this ensemble imposes a *structure* upon an existing reality, and it is the combination of structure and the *situation* as determined by physical and biological facts that determine individual incentives and opportunities. According to Schmid, economists have naively assumed that an individual's behavior is shaped merely by preference rankings of exchangeable goods, failing to examine how shared norms and public policy shape the opportunities for choice.

For purposes here, Schmid's framework will be summarized in terms of four key elements. They are defined here with explicit attention to the analysis of food safety and nutrition policy.

1. *Situation*: the things that cannot be changed. This should be understood to include the physical, chemical or biological processes that determine the production and consumption of foods.
2. *Structure*: the ensemble of laws, shared norms, procedures, and rules that are either proposed or in place in the status quo. In addition to the obvious elements of policy, structure includes norms that govern what people regard as food.
3. *Conduct*: the behavior that will be produced as a result of the opportunities created when a given structure is imposed upon the situation. Conduct includes the production, processing distribution and consumption of food.
4. *Performance*: a given pattern of conduct will produce an end state which consists of the policy's consequences for affected parties. Health, disease, injury, profit and loss all qualify as components of this end state. (Schmid, Shaffer and van Ravenswaay, 1983)

The framework provides general categories that allow a competent analyst to bring implicit features of policy out more clearly and to examine how policies produce end states. It is admittedly quite general and is undoubtedly commensurate with many different methods of policy analysis.

It is worth noting a few additional points before examining how ethics bears upon the framework. First, the framework is interpretive in that it will require judgment to assign specific variables to any of the four elements. For example, the technology that is used to detect the presence of a substance in foods utilizes physical and chemical principles. Technology is, in that sense, a part of the reality or situation on which a structure is imposed. However, this technology has changed so dramatically in the past four decades that it is probably more useful to think of it as a component of structure. The interpretation of administrative guidelines for food safety decisions includes standard operating procedures for the use of specific technological tests. As such, when technology changes, there is a sense in which policy changes, too.

Second, the general category of performance can be taken to include the full range of criteria that would be applied in evaluating a policy. As will become clear shortly, some such criteria have little to do with the end state produced by the policy. The dominant practice in public policy analysis is to predict policy outcomes and to report them as an end state, often as costs and benefits. This practice leaves the decision to the responsible party or parties, be they an administrator, a court or the Congress. Decision makers can and do apply criteria that make little if any use of predicted end states, but the typical practice among analysts is to equate the predicted end state with the policy's performance. Analysts writing on the banning of Alar, for example, typically evaluate the policy in terms of a trade-off between the economic value of the apple crop for producers and some minimal, even tentative, reduction in risk for consumers. This approach leaves open the possibility of comparing these outcomes according to a variety of criteria but presumes that the decision is either based upon projected policy consequences to the extent that it is defensible at all. (Roberts and van Ravenswaay, 1989).

How Ethics Bears on Policy

The assumption that consequences (or end states) provide the basis for evaluating public policy has its philosophical basis in the ethical writings of Jeremy Bentham and John Stuart Mill. These utilitarian philosophers argued that action can be justified only in light of the consequences, and they proposed the twin norm of counting consequences for all affected parties and of maximizing aggregate utility. Traditionally utilitarian philosophy has been criticized for its insensitivity to the distribution of costs and benefits. In more recent times, John Rawls (1971) has argued that policies should benefit the worst-off groups, rather than maximize aggregate utility. As Nozick (1974) noted, both utilitarian and Rawlsian egalitarian theories evaluate policy by applying a norm or decision rule to the end state that the structure is expected to produce. Many economists who do not think of themselves as either utilitarian or egalitarian would also assume that end states provide the sole basis for evaluating policy. The search for Pareto better solutions or efficient levels of pollution begins by predicting policy outcomes. The debate consists of whether the accounting is accurate and complete *or* which principles to apply in evaluating end states.

To use the language developed here, these are all *performance* or *end state focused* approaches to ethics. Their philosophical pedigree extends back to Bentham, who hoped to reform an English legal system based upon status and privilege. By turning the debate toward consequences, Bentham established a burden of proof for which social rank and divine right were irrelevant. While common law based policy on past practice, Bentham's theory held that it *should* be based entirely on expected outcomes. Bentham assumed that rationality consisted in actions chosen as means to an end. While one might disagree about ends, Bentham thought that a rational person must accept that acts which fail to achieve the desired end are to be rejected. The point to note here is that this makes structure into a means toward an end. While citizens in a democracy can be expected to have different preferences, this view entails that the rules and regulations adopted by policy makers can only be evaluated as means to *some* end. Hence end states are the dominant performance criteria for public policy.

Nozick's *Anarchy State and Utopia* (1974) is an extended philosophical attack upon this notion of political rationality. He offered a now-famous analysis of why the basketball player Wilt Chamberlin is entitled to great wealth, despite what Nozick thought to be the lack of any proportionate social value produced by his play. The argument stressed that given any initial distribution of wealth, policies that confiscate money voluntarily exchanged between Chamberlin and paying fans must necessarily violate individual liberties. While a performance focused analyst might be able to argue that the market structure permitting such exchanges is efficient (in that it maximizes utility or produces a Pareto better outcome), such considerations are irrelevant for Nozick's argument. The point was that government policies are justified *only* when they conform to an antecedently determined set of moral or political rights. The consequences produced by structures conforming (or failing to conform) to this set are irrelevant.

Just as it is possible to differ over the performance criteria used to evaluate policy, it is possible to disagree about which rights belong in the antecedently determined template used to evaluate policy. For Nozick and other libertarians the template of rights will be narrowly confined to those which protect individuals from interference by others. For liberals such as Ronald Dworkin (1977) or Henry Shue (1980), the list of rights may be much more expansive. The point here is that these approaches to policy are *structure focused.* They establish a burden of proof met only by (1) demonstrating that policy conforms to the antecedently chosen structural template; or (2) refuting the claim that a given right belongs in the template. Arguments that stress trade-offs, efficiency or other features of end states do not meet either test.

The tension between end state and structure focused policy evaluation has a ready analogy in the debate over food labels. Whether required by law or custom, labels are clearly a component of structure for food policy. The debate is: how should labels be evaluated? Given a performance focus, labels will be evaluated in terms of the end state they produce for the producers and consumers of food products. Labels will be seen as educational tools. One will want to know whether the label allows the consumer to make food purchases that more fully satisfy preferences. Policy will be seen as a trade-off between producer costs, consumer preferences, and health, and labels can affect each of these outcomes in a variety of ways. A structure focused evaluation will see the matter almost entirely in terms of informed consent. Labels will be seen as transforming the conditions of consumer choice from those of mild coercion to implied consent. Policies that protect individual consent are acceptable; those which foreclose individual consent must bear a very heavy burden of proof before being judged acceptable. Labels will be preferred even if people choose to ignore them, or even if consumers have false beliefs that lead them to make less than optimal choices.

Utilitarian or consequentialist ethical arguments, then, express norms or policy criteria that focus on end-state performance, while human rights arguments identify characteristics of structure that must be in place without regard to consequences. The matter does not end there, however. It is often a person's conduct that is judged ethical or unethical. If a policy structure induces individuals to behave in ways that are unethical, there is a basis for rejecting the policy. For example, a pesticide policy which encourages producers to misrepresent their use of chemicals would be judged unethical, even if no rights are violated, or if no harm is done.

Many authors have taken up conduct-focused ethics in the past two decades. Bernard Williams (Smart and Williams, 1973) criticized utilitarian arguments because they fail to address the character of the moral agent. Alisdair MacIntyre (1981) has criticized both rights theory and utilitarian arguments for the emphasis that they place on an individual's self-regarding wants. He argues that a better approach would take up virtues and vices that are the reference points for moral character. These authors do not discuss policy, however. The public choice approach to policy analysis makes it possible to see how these philosophical ideas bear on policy by making it clear how situation and structure produce conduct. There is little doubt that most citizens would regard the conduct focus as the most obviously "ethical," despite the relative lack of attention it has received by policy analysts. Public outrage over congressional check bouncing, for example, is almost certainly focused upon conduct, rather than structure or performance.

With respect to food safety policy, the most likely relevance of conduct-focused evaluation is not to producers and consumers but to the conduct of policy makers themselves. Arguments against the practice of pricing life, for example, are best analyzed as an objection to the practice of quantifying the value of life. Annette Baier (1986), Allan Gibbard (1986), and Douglas MacLean (1990) have expressed the concern with pricing in life in this way. The problem cannot be resolved by adjusting the amount of value assigned to lives, nor is the objection based on the suggestion that human lives should be assigned infinite value. The point is that persons of good character do not make decisions by attempting to decide how much others are worth. A policy procedure which requires public servants to engage in such conduct is, in this view, a corrupt and indefensible procedure, and the policies that result from it are tainted.

To sum up, ethics bears on policy in three ways. Traditional human rights arguments focus upon structure, applying a template of antecedently determined constraints in their assessment of policy. The language of virtue and integrity focuses on conduct, evaluating a policy in terms of the patterns of behavior it promotes. Finally, the end state or performance evaluation of policy that has become a staple of the social sciences draws upon the utilitarian tradition in ethics. Each approach establishes a burden of proof that cannot be met by argu-

Table 7.1. How Ethics Bears on Policy

FRAMEWORK	ETHICS
Situation Structure	Rights Theory
Conduct	Virtue Theory
Performance	Utilitarian Theory

ments grounded in either of the other two approaches. At the same time, each approach is deeply grounded in the culture and habits of contemporary Americans.

Food Biotechnology, Policy, and Ethics

An analysis of food labels that stresses rights and consent will establish very different burdens of proof than does an analysis that evaluates labels according to their consequences. Matters of character and conduct enter only indirectly into the disputed areas of policies for consumer information but could be decisive to the extent that they break a deadlock between those focused on structure and those focused on performance or outcome criteria. Recombinant bovine somatotropin (BST) is the case that has spurred debate. The substantive ethical issues raised by the development and proposed release of BST concern unintended consequences for the dairy industry, dairy animals, and for environmental impact of dairying. However, it is public acceptability of milk produced using the new technology that has produced the greatest anxiety (see Chapter 9). Opponents of BST have called for a ban on the technology and, short of that, for labeling of milk produced using recombinant BST (Hanson, 1991). The ethical evaluation of labeling policies for food biotechnology has therefore already assumed practical importance (see also Hopkins, Goldberg and Hirsch, 1991).

The case of BST should serve as motivation for thinking about ethical issues, but it is a poor example of the scientific issues that will arise in connection with the use of biotechnology for food products. At least some of these products will pose difficult questions for risk analysis, and some may pose quantifiable risks. By contrast, there is virtually no scientific support for questioning the safety of BST milk. (Kroger, 1992) Future policy decisions will almost certainly be characterized by the kinds of uncertainty that have hindered the application of science to public policy in the regulation of artificial sweeteners (Merrill and Taylor, 1986), or of chemicals (Graham, Green and Roberts, 1988). In such cases, questions about the extrapolation of data from animal studies, or about the applicability of epidemiological data caused regulatory policy to become embroiled in technical and methodological disputes. Criteria for scientific judgment and cross disciplinary conflict over patterns of scientific inference are crucial to the policy debate. Because the procedures and norms for scientific inquiry are themselves matters of philosophical dispute, it is accurate to describe risk policy debates as philosophical controversies (Hollander, 1991). Debates over acceptable evidence extend philosophical controversy into the interpretation of situation, of the basic facts that must be accepted as constraints upon available policy options. These debates are not, however, ethical debates that conform to the pattern described above.

Although the debate over acceptable evidence will almost certainly recur in future food policy decisions, the lack of scientific or technical controversy over BST makes it a good example for considering ethical issues precisely because disagreement about the probability of harm does not confuse the ethical issues. The biological facts that comprise the situation for BST milk are not themselves a source of controversy, at least not among scientifically informed participants. Nevertheless, labeling requirements for BST milk have been proposed. A policy that certified or required labels should be understood as an alternative to policies which regulate by removing or approving products *tout court*. Labels thus become a component of policy structure, to use the terminology introduced here. It is worth noting, however, that labels might become a component of structure in any of several ways. One might require labels that proclaim the presence of BST milk, or one might permit the use of labels that certify its absence. In either case, the precise wording of labels will be extremely important, as will the procedures for assuring the integrity of labels. The diversity of approaches to labeling implies that it is not one policy proposal that is being discussed here, but a general class of potential policies that would be evaluated in similar ways.

Whatever labeling strategy is employed, a policy utilizing product labels can be expected to stimulate certain patterns of conduct by consumers and by processors and manufacturers. Some consumers will read labels and will use information as a basis for food purchases; others will not. One would presume that consumers expressing concern over BST in milk would use the label, while others might not. These patterns of conduct will lead to consequences that determine the performance of a labeling policy. Relevant consequences certainly include health benefits or risks to food consumers that are incurred on the basis of their conduct. They also include costs to consumers in the form of higher food prices, and in the trouble and inconvenience required for reading labels. Costs to processors and manufacturers are also a component of the policy's performance. Given this account of situation, structure, conduct and performance, it is possible to examine the ethics of a labeling policy for BST.

Performance Focused Evaluation of Labels: As noted, product labels can be expected to produce certain costs and benefits for consumers and producers. In the case of BST, the scientific consensus is that the health benefit to a person who would use such a label is zero. Consumers who express concern over the use of BST would derive some benefit from reduced anxiety if they are able to satisfy their preference for non-BST milk. These benefits must be weighed against the direct costs of labeling, costs which may be significant when their impact upon processing is assessed. Even the approximate value of these costs and benefits is largely speculative, but the point here is to see how consequence assessment provides an ethical basis for the evaluation of policy. The policy is justified in terms of the acceptability and desirability of its consequences. The historical standard has been the utilitarian maxim proposed by Jeremy Bentham in 1789: act so as to produce the greatest good for the greater number of people. Although there are many cases in which pure optimization rules such as the utilitarian maxim may need to be modified (Thompson and Stout, 1991), policies which do not provide benefits that compensate for their direct cost of implementation to government and to affected industries will always be difficult to justify.

Structure Focused Evaluation of Labels: Structure focused evaluation centers upon protection of rights as a precondition to ethically legitimate application of state power. Two key criteria are consent and fairness. The principle of government by consent of the governed is, in many respects, the foundational norm of democratic government, while fairness, understood as equality before the law and protection of minority rights, constrains the excesses of democratic decision making. Labels are an attractive component of structure because they make it possible to argue that individual food consumers have been placed in a position to grant or withhold consent to food-borne risks (real or alleged). A consumer who chooses to purchase a labeled food item can legitimately be understood to have consented to the transaction, so long as meaningful alternatives are available. A policy structure which does not allow consumers to discriminate on criteria they have judged to be important violates consent criteria. However, labels can also raise questions about fairness to the food industry. If the institutional practice is to use labels only in cases where serious risks to health have been scientifically demonstrated, as has been the case for tobacco, then the application of a label to BST milk may violate the rights of industry by unfairly prejudicing consumers against the product.

Conduct Focused Evaluation of Labels: Character and virtue are less clearly related to labeling policy than are rights and consequences. It is not obvious how the presence or absence of labels for BST milk would induce consumers to engage in unethical conduct. Some Americans may be inclined to make moral judgments of character based upon a person's dietary choices. Some religions require a dietary regimen for the devout, for example, and it is already common for vegetarians and non-vegetarians to make moral judgments about one another. Even so, there is little public consensus for such judgment. The more relevant conduct is that of industry. To the extent that labels represent a form of disclosure that would be required by norms of honesty or truth-telling, a practice of labeling might be thought to promote ethical conduct on the part of industry. Ironically, disclosure will win far more praise if it is voluntary. Hence, a labeling policy will win more praise from those who focus on conduct if it facilitates, but does not require, disclosure of relevant information. Conduct evaluation does not provide a clear mandate for or against labels, however, and it will not be emphasized in the comparison which follows.

Comparing Ethical Approaches for Policy Evaluation

Structure focused and performance focused approaches to policy evaluation establish different and sometimes contradictory burdens of proof. Evidence and argument which is highly relevant to a performance evaluation is often irrelevant to an evaluation in terms of rights and consent.

There are a number of difficult philosophical and economic measurement problems that must be addressed within a performance focused evaluation (Giere, 1991). Two general points of dispute concern the quality of models used to predict health consequences and the choice of a decision rule to compare consequences, once they have been assessed. The debate over linear vs. threshold extrapolation of data (Schaffner, 1991) is an example of the first problem. The debate over Delaney vs. *de minimus* is an example of the second (Jasanoff, 1991). The point here is to see that the crucial burdens of proof established within a consequence evaluation approach differ from those in a rights-based procedure. Within a structure focused evaluation, consumer choice is important to the extent that it satisfies criteria of consent; questions of whether consumers are made better off by the choices they make are irrelevant.

Rights-based approaches to social theory have never assumed that governments have any responsibility to make socially optimal policy decisions. Rather, the first ethical responsibility of government is a negative one: *not* to interfere in the personal liberties or freedoms of its citizens. Accordingly, structure focused evaluation of public policy stipulates a list or template of rights, liberties, and possibly opportunities, much like the United States Bill of Rights. A policy is evaluated in terms of its ability to satisfy or fit this antecedently determined template of rights. The list of rights is adopted prior to entertaining any particular policy option, so performance evaluation of specific policies is not a component in justifying the inclusion of a given right. The key right with respect to food consumption is a general right to non-interference in personal choices, provided that personal choices do not violate complementary non-interference rights of others. The example of labels for BST provides an illustration of why performance focused and structure focused approaches to understanding how ethics bears on policy introduce distinct burdens of proof. A more detailed analysis of labels would need to take up additional issues. For example, the traditional norm of *caveat emptor* has historically served as an informal component of structure for consumer food decisions. The previous discussion of labels has assumed that consumers have a right to any information they deem relevant about food choices, but caveat emptor might be taken to qualify this right. Further discussion of issues specific to labeling policies cannot be undertaken within the constraints of this paper. However, readers should be advised that what has been said to illustrate the contrast between structure and performance is far from being a complete ethical analysis.

The contrasting burdens of proof for performance philosophies and structure philosophies present a general problem for biotechnology policy and food safety. Interested parties, including scientists and regulators, can be so closely wedded to one of these philosophies that they fail to understand the force of their opponents' arguments. Someone who insists upon interpreting structure-focused rights arguments in terms of the end state produced by policy will simply miss the point of that argument. Rights arguments will appear as irrational or naive, failing to grasp the importance of trade-offs that are thought to be the main focus of policy evaluation. It is not difficult to find authors who appear to exhibit virtually total insensitivity to the burdens of proof entailed by a focus on structure. In a 1990 article on food safety policy for recombinant DNA derived animal growth hormones, Fred Kuchler, John McClelland and Susan Offutt (1990) characterize the issue entirely in terms of performance or end-state criteria. The idea that rights or consent are relevant is absent from their analysis. Doyle and Marth (1991) also offer an analysis focused exclusively on end states.

Milton Russell (1990) has written that it is irresponsible for a public official to make policy decisions without attempting to assess the consequences to the fullest extent possible. He qualifies this commitment to performance evaluation by also stating that "legitimacy flows from an acceptance of the decision, or at least the decision process, by those affected." (p 22) This qualification stresses the importance of consent. Consent *can* involve the prediction of consequences, so long as predictions are part of an information sharing process designed to build consensus. But traditional consent criteria stress the protection of rights and may leave the burden of predicting consequences to the affected parties themselves. Despite his qualifying comment, Russell may favor performance criteria in making the prediction of consequences a strict requirement of ethical policy making, even while he endorses the principle of consent. One aspect of the tension between end state evaluation and principles of consent is the question of who assesses consequences.

Biologists or economists who predict the consequences of biotechnology do not, generally, need the advice of citizens in the process of collecting data and making projections. They can produce a prediction of end states without seriously involving citizens in their activity. The key predictions for regulation of foods involve risk assessment and economic impact. If citizens are invited to participate in decision making after such information has been collected, analyzed and presented, the opportunity for a structure or conduct focused evaluation has been reduced, if not foreclosed, by the preponderance of evidence relevant only to performance criteria. Since structure focused criteria often stress the importance of participation and consent, the lack of citizen participation in the scientific assessment of risk and of economic impact is doubling troubling. Citizens have been denied participation in the early stages of the process and are faced with a final decision procedure in which evidence that is potentially irrelevant to consent criteria dominates.

Structure and conduct criteria should not be arbitrarily excluded from a decision, as they are when scientists make risk assessments without substantial citizen participation. But this is *not* to say that citizen assessment of risks should simply be substituted for scientific ones. Frank Cross (1992) accuses me of doing just this in a 1990 paper

(Thompson, 1990c). My point is that end state assessments cannot meet burdens of proof established by criteria that stress participation and consent.[1] People who interpret the criticism of scientific risk assessment as a call for citizen assessment are displaying a myopic focus upon performance criteria. Structure and conduct criteria establish burdens of proof in which *anyone's* assessment of likely end states is largely irrelevant. Put another way, if the goal is to implement a policy that is likely to minimize food related illness, then it seems obvious that the best scientific techniques should be used to predict the incidence of illness associated with a given product or practice. However, if the goal is to ensure that consumers have confidence in the food supply, an altogether different policy may be indicated. Consumer confidence may be very imperfectly correlated with the probability of illness. Confidence may be more closely correlated with participation and consent, with the structure under which dietary decisions are made rather than the end state that is produced.

The implication is that, while performance evaluation is "objective" in the sense that it does not systematically favor any specific interests or political ideology, the practice of performing extensive scientific studies can introduce a bias against ethical criteria that emphasize structure and conduct. Yet it is easy to find examples of structure or conduct focused criteria in the history of American government. The framers of the U.S. Constitution were themselves structure-focused in adopting the Bill of Rights and conduct-focused in proposing the division of powers. It is therefore reasonable to presume that democratic decision making ought not be systematically biased against structure and conduct criteria. Baring specific arguments to the contrary, policy-making procedures should avoid domination by any one of these three philosophical approaches to ethics and should weigh evidence and arguments in terms of the burden of proof to which they are most clearly relevant.

---

1. Deborah G. Mayo (1991) has an extended discussion of how scientific and citizen assessments of risks might be compared. Thompson (1997a) includes a longer analysis of the issues discussed in this chapter.

8

Animal Welfare, Animal Rights, and Agricultural Biotechnology

Animal well-being issues relate to biotechnology policy and to ethical issues in biotechnology in four ways. First, criticisms of the use of animals in scientific research have led to reforms in and constraints on research practice in all areas of the biological sciences. As such, issues of animal well-being are having an increasing impact on the biological research agenda, and upon the costs and conduct of research in which animal models are deemed useful.

Second, the fact that there are potential political allies for opposition to biotechnology may lead animal activists to target the products of recombinant DNA research for concerted attention. In the case of recombinant bovine somatotropin (rBST), animal activists aligned with dairy farmers and with groups representing small dairies in Northern states. Since there is already some degree of elevated public concern over products derived through recombinant DNA, virtually any new product for animal agriculture will be a candidate for scrutiny by individuals organized to speak for animal interests. Papers by Jeffrey Burkhardt (1991) and Dorothy Nelkin (1991) each noted that groups and individuals acting on behalf of animal interests have had an impact upon public receptivity to rBST.

Third, the potential for using recombinant DNA techniques to develop new animal genomes may raise special questions for animal well-being. Will transgenic animals be vulnerable to physiological or psychological dysfunctions that are either qualitatively different from pathologies associated with conventional animal breeding, or will they experience familiar problems at a higher rate? Are existing animal care procedures, including procedures for euthanization of dysfunctional individuals, adequate? Does our understanding of well-being include entitlements to certain key functional traits such that it would be unethical to select against these traits? These and other questions like them have not previously received focused attention; they will come to the forefront as animal biotechnology moves forward. (Thompson, 1997b)

Finally, the public's perception of a lack of compassion for animal interests on the part of the research community may be a component of a vague, but extremely significant, antipathy toward science in general. Opposition to biotechnology may be cited along with recent reports of scientific dishonesty or padding of indirect cost recovery as yet another aspect of public disenchantment with science and scientists. If so, a willingness to understand and take seriously the issues of animal well-being and of public attitudes toward animal interests will be a component of responsible science in the coming decades.

Discussions of ethical issues in the care and use of animals commonly assume a distinction between animal welfare and animal rights; but the distinction itself has become the source of much confusion. There are, in fact, at least three ways to draw the distinction referring to political, conceptual and philosophical objectives, respectively. There is virtually no correspondence between these three ways of making the welfare/rights distinction. Use of the term "animal rights" at the level of a conceptual distinction, for example, does not commit one to advocacy of animal rights at the political or philosophical level. All three levels of analysis are in frequent use during discussion of animal ethics, however, and it has become essential for informed participants to understand the three levels on which the distinction can be made. Each type of distinction is discussed below, and the relevance to biotechnology is noted.

The Political Distinction

Among groups active in promoting legislation, reform, and general awareness of animal issues, those who advocate relatively moderate reforms describe themselves as animal welfare activists, while those who advocate more sweeping reforms describe themselves as animal rights activists. The United States has a long history of commitment to humane treatment for animals. This commitment is reflected in existing humane treatment laws and in the existence of individuals and organizations that are committed to the well-being of all animals. These people are commonly described as advocates of animal welfare. More recently, individuals and organizations have begun to work for change in social attitudes toward animals and have promoted policies that severely restrict humans' use of animals. These people are described as advocates of animal rights.

While their main efforts on behalf of animals focus on voluntary care and enforcement of existing laws, individuals and organizations in the animal welfare camp remain committed to changes in laws protecting animals. Some, for example, oppose hunting and trapping practices that are judged to be cruel. They have also opposed the use of animals in certain forms of product testing, notably the Draize test (where caustic and non-caustic substances are applied to one eye of a live rabbit). When they judge that tests do not provide significant advances in knowledge or other benefits for humans animal welfare activists become quite vehement. Nevertheless, the political agenda of animal welfare advocates is, with respect to biotechnology at least, relatively moderate. For the time being, it is reasonable to expect that research in medical biotechnology promises improvements in human health that will likely justify the use of laboratory animals in the minds of animal welfare advocates.

In the past two decades, a new type of individual and group has emerged. The new groups are committed to sweeping change in the use of animals, and, while not universally committed to vegetarianism as a moral principle, they are committed to the view that use of animals for food should be far more limited than is currently the case in industrialized countries. These people are commonly described as advocates of animal rights. The political and cultural roots of the animal rights movement are complex. The movement includes some individuals who, in decades past, would have been part of the animal welfare community, but it also includes people who see animal rights as an extension of political reforms that extend rights and political power to women and minority groups.

To date, animal rights activists have focused more attention upon animals used in medical research than upon agricultural animals. As such, their impact upon biotechnology research may come in the form of opposition to research in molecular genetics that utilizes research animals. At present, however, these groups are increasingly turning their attention to agriculture. They may be forming alliances with environmental causes and with long-standing opponents of biotechnology. Jeremy Rifkin's 1992 book *Beyond Beef,* for example, is an extended attack upon the United States cattle industry. If such an alliance develops, there will be as much or more attention paid to agricultural biotechnology as to the use of animals in medical biotechnology research.

People for the Ethical Treatment of Animals (PETA) and the Farm Animal Reform Movement (FARM) are animal rights organizations. Groups such as the American Society for Prevention of Cruelty to Animals or the American Humane Society are self-described animal welfare organizations. The political agendas sought by various individuals and groups are extremely broad and divergent and cannot be reviewed further here. As already noted, when opportunities for political success are stimulated by bursts of media coverage and by events such as the controversy over recombinant bovine somatotropin, both welfare and rights groups can be expected to take an interest in biotechnology.

The Conceptual Distinction

The terms "welfare" and "rights" have important meanings for law, economics, ethics, and political theory, but their meanings vary by context. *Welfare* usually means well-being, referring to a subject's relative state of health, happiness, or satisfaction. But welfare is also used to indicate entitlement programs intended to benefit indigent or disadvantaged persons. Measurement of welfare or well-being is a contentious issue in terms of whether farm animals or human beings are the focus of concern. Once a measure of welfare has been agreed upon, the goal of welfare-oriented policies is to identify options that maximize or at least optimize total well-being to all affected parties. Since welfare can rarely be achieved without cost, a key consideration in achieving welfare objectives is the ratio of benefit to cost, or the *efficiency* of policy. An option that increases well-being, but at greater cost than the alternative, is inefficient. Criteria of efficiency become vexed when (as in reality) it is costs to one group that must be weighed against benefits to another. In the case of animal welfare, it is trade-offs between human and animal well-being that pose ethical dilemmas.

It is multiple uses of the term *rights,* however, that create the most serious confusion. Joel Feinberg (1970) has analyzed the core meaning of *rights* in terms of valid claims that can be made by or on behalf of one party against another. That is, one would have a right to something if and only if one could make a valid claim to it. The criteria for validating claims depend upon the frame of reference. If the claim can be validated through criminal or civil actions in a court of law, the claim represents a *legal right.* If the claim is validated by general principles of morality and ethics, the claim represents a *moral right.* There can also be informal rights validated by custom or by rules of etiquette. Although there are clearly areas of overlap among law, ethics and custom, the existence of a right to something within one frame of reference does not guarantee the existence of a right to that thing in another.

For example, custom confers upon the first person in line the right to be served next. Custom also confers a right to their place in line to others who wait. While these rights will be recognized by almost everyone who shares the custom of queuing, it is unlikely that people have legal rights to their places in most queues. Similarly, while it might be mildly unethical to flagrantly violate the custom of queuing up for service, the fact that some societies do not recognize this custom is a reason to doubt that there is anything resembling a moral right to one's place in line. It is also worth noting that this informal right is highly contingent and context bound; it does not entail a series of supporting rights. That is, one does not have a right to goods such as food, health care, or sanitary facilities that, under easily imaginable circumstances, might be necessary to guarantee one's right to a place in line. There are instances in law and morality where having a right becomes meaningful only when a series of additional rights are granted, but it is not part of the concept of a right that such supporting rights are also implied.

The contingency of rights is especially important for our understanding of biotechnology issues. It is clearly meaningful to make claims on behalf of animals, and likely that some claims made on behalf of animals would be validated by custom at least, if not also by law and ethics. Americans are in virtually unanimous agreement that animals should not be subjected to sadistic and intentionally cruel treatment. A person who intervenes on behalf of a cruelly victimized animal is making a claim on behalf of the animal that is validated by custom, ethics, and, in all likelihood, by law. It is therefore meaningful to assert that the animal has a right not to be treated in a cruel and sadistic manner. Asserted in this restricted context, this right does not imply supporting rights such as a right not to be killed and used for food or fiber. One may speak of certain limited animal rights without implying a commitment to the extensive political changes sought by so-called animal rights organizations.

In addition to implying that a valid claim can be made, the concept of rights is usually used to assert the priority of an individual or minority interest over aggregate interests, powerful interests, and even the common good. In recent American politics, this use of the term *right* signals the rejection of an evaluation process that stresses "tradeoffs" or optimizing the ratio of benefit to cost. A common example is found in property rights. When deciding how to dispose of Jack's fifty dollars, Jack's neighbors might consider how several alternatives would most benefit them as a group, or what expenditure would be in the public interest. These factors are not considered, however, because

Jack is recognized to have the right to use his fifty dollars however he chooses. The fact that his choice takes no account of trade-offs to the interests of the larger group, of powerful persons, or even of the common good, is not thought to establish a valid claim against him.

Ronald Dworkin (1977) has introduced the idea of "trumps" to explain this aspect of rights. While trade-offs may be made in some areas of policy, the process of examining costs and benefits is constrained by rights. Trades that would violate individual rights are "trumped" by valid rights claims. Contentious policy issues often revolve around where the line between trades and trumps is to be drawn. In natural resource policy, for example, there has been an enduring controversy between those who wish to weigh the benefits and costs of alternative policies and those who assert that endangered species should be protected even when it is not cost efficient to do so. This latter group can be said to be making a rights claim on behalf of the endangered species, while the former group might be described as being concerned with the general welfare.

The contrast between trades and trumps illustrates how *welfare* and *rights* can be opposing concepts in a variety of legal, political and ethical controversies. Those who feel that certain individual interests are more important than the benefits sacrificed to protect them will be attracted to the language of rights for expressing their objectives. Those who feel that there are no valid reasons to sacrifice benefits will stress the general welfare or will stress that efficiency is the key policy goal. Those who feel that lower prices for consumers or higher profits for producers are more important than certain perceived animal interests are inclined to stress an aggregate or common good. Those who feel that important animal interests are about to be sacrificed will try to "trump" that efficiency argument by claiming that an animal's rights have been violated.

The Welfare/Rights Distinction in Philosophical Ethics

The preceding section describes the conceptual distinction between rights and welfare, but the justification of moral rights has not been considered. A moral right has been understood as a claim validated by ethical principles. What are the ethical principles that would validate such a claim? There are at least two different types of principle that might be relevant. First, it might be the case that one will bring about the best consequences in the long run by recognizing certain claims made on behalf of individuals, even if it is inefficient to do so in the short run, or in a particular case. We might recognize a person's right to a place in line because this convention protects social order, even though there are specific instances where a person's place in line is a trivial benefit relative to costs imposed upon others. Rights claims produce a social agreement that determines the rules of the game. Having a stable set of rules can reduce long-run costs enormously, as we can plan our affairs with the knowledge that, though we must recognize certain claims made against others, we may make claims against others ourselves. It would also be quite costly to actively calculate costs and benefits for each social transaction. A good set of rules, some stated as rights, will, therefore, promote efficiency. In this rationale for rights, the ethical principles which validate rights claims appeal to the long-run good, or to a more sophisticated notion of social efficiency.

The second type of rationale stresses the absolute moral importance of the individual. Here it is individuals who are understood to be the focus of morality, and it is our respect for another individual's personhood and autonomy that is promoted as the central idea in ethical action. In this view, there are ethically deep traits of individuals that must be respected and protected if the fundamental terms of ethics are to be met. At least some of these traits are rights, sometimes called natural rights, human rights, or metaphysical rights. In this view, our responsibility as individuals is to respect others' rights, and our responsibility as a society is to adopt codes that support morality by extending police and legal protection to these crucial rights. The religious and philosophical rationales that support this second way of understanding rights vary. Some base their vision of human nature on theology, others on rationality, and still others on assumptions implied by the coherence of moral language and political rights. The elements of these philosophical positions will not be discussed here.

The two distinct rationales for showing how a rights claim might be validated by ethical principles stipulate very different kinds of ethical principle. The first takes an action or policy to be justified in light of the consequences it produces. Philosophers such as Jeremy Bentham and John Stuart Mill argued that all ethical principles were ultimately reducible to a principle called the *utilitarian maxim*: act to produce the greatest good for the greatest number of people. In Bentham's development of the principle, ethical choice was described as calculation in which the benefits and harms succeeding an action were first assessed for each affected party and then summed for all affected parties. The right action was taken to be the one that produced the most total benefit or the least total harm. Beginning with Mill, philosophers have noted that Bentham's picture of this calculation is at least too simple, and have proposed more-sophisticated ways to measure and assess consequences. The utilitarian tradition makes welfare considerations philosophically fundamental and justifies rights claims in terms of impact upon general welfare. Thinkers within the utilitarian tradition have remained committed to the idea that consequences for all affected parties must be weighed in the calculation, and that benefits and harms (or now costs) are the rough units in which consequences are to be measured.

The second strategy has deep roots and clearly inspired the framers of the U.S. Constitution to include a Bill of Rights. It can be found in the writings of Aquinas, Locke and Rousseau, but more recent advocates include Gewirth (1982) and Rawls (1971). Here, rights are the fundamental philosophical concept. As noted, the rights view takes it that there are traits—rights—that must be protected or guaranteed, and that the morality of an act is to be judged according to whether it successfully respects the rights of others. The dual implication of this approach is that some acts judged moral by utilitarians in virtue of producing the greatest good will be judged immoral by rights theorists when individual rights are sacrificed, while some acts that are clearly inefficient when judged by the utilitarian standard are fully consistent with the terms of morality laid down by rights theory. As such, there is a deep philosophical tension between utilitarian philosophers and those who have constructed moral theories based on a concept of rights.

Ethics and Biotechnology

There are three closing points to be made with respect to ethics and biotechnology. First, the distinction between welfare and rights extends into the deepest levels of philosophy, but there is no necessary correspondence between *philosophical commitments* to welfare or rights and practical, *conceptual commitments* to the assertion or denial of specific rights claims, nor between either of these and the *political commitment* to groups organized around animal welfare or animal rights objectives. The logical and casual links between philosophical views and political activism are contingent upon other factors which make ethical views a poor predictor of an individual's opinion on biotechnology. Second, although there are different philosophical beliefs and traditions to support rights philosophy, they converge on the belief that philosophies which fail to recognize the primacy of the individual over the general good abandon the most fundamental principles of ethics. As such, those who are philosophically committed to animal rights will conclude that social benefits from biotechnology are "trumped" by harms to individual animals. Third, major figures in radical (e.g. animal rights) political organizations differ over which deeper philosophical principles best justify the radical initiatives on which they agree. These differences present opportunities for activists and biotechnologists to engage in more-sophisticated debate at the philosophical level than has hitherto taken place. Each of these three points is summarized below.

Discussion of animal welfare and animal rights is confusing because the terms *welfare* and *rights* might refer to the deep philosophical tension between fundamentally opposing approaches to ethics, but they might refer to the more superficial distinctions already discussed. As just noted, no firm correspondence holds across levels. A utilitarian may well conclude that establishing a legal or custom right is the most efficient way to promote the greater good. A rights theorist who feels that no fundamental rights are at stake with animals may promote a welfare approach. There is certainly no correspondence between the philosophical and political levels, as some of the most radical activists are utilitarian (e.g. welfare-oriented) philosophers, while many rights theorists resist the extension of philosophical rights claims to non-humans.

The potential for confusion is multiplied by the fact that there are several different philosophical theories that are often included under the rights banner. For the purpose of understanding animal rights views, however, the differences between these views are less important than the fact that they share a rejection of utilitarian emphasis upon making trade-offs between costs and benefits, at least where key rights are concerned. This point is made clear by Tom Regan in an article entitled "The Case for Animal Rights" (1985)[1]

> What has value for the utilitarian is the satisfaction of an individual's interests, not the individual whose interests they are. A universe in which you satisfy your desire for water, food and warmth is, other things being equal, better than a universe in which these desires are frustrated. And the same is true in the case of an animal with similar desires. But neither you nor the animal have any value in your own right. Only your feelings do. (p. 19)

---

1. This article from *In Defense of Animals,* edited by Peter Singer, is not to be confused with the book by Regan entitled *The Case For Animal Rights* (1983), which contains an extensive discussion of animal behavior and of how to balance conflicting rights, in addition to a rejection of utilitarianism.

Regan goes on to criticize this view with a story in which the rich but stingy Aunt Bea is murdered and her wealth is distributed to needy people. Regan adopts the rights view because he thinks that utilitarianism justifies this act in virtue of the greater good achieved. Although this is a very incomplete argument for rights, it is a conclusion widely shared by rights theorists, including those who are unwilling to extend rights to animals. All those who argue philosophically for animal rights reject utilitarian ethics.

The utilitarians themselves have extended moral concern to non-human animals for the simple reason that non-humans have feelings, too. Non-humans experience pain and satisfaction, though the character and degree of these feelings can be difficult to assess. For example, Peter Singer's well-known work on animals derives from the simple observation that some animals, including most agricultural species, undoubtedly feel pain. Singer's utilitarian views lead him to conclude that the suffering of non-human animals should be weighed against the benefits that humans derive form the use of animals. Singer's philosophical work on animal suffering denies the validity of animal rights, except in so far as rights claims are based on underlying consideration of tradeoffs. Singer does not think that animals have a right to life and, indeed, has argued that humane slaughter of an animal is not a serious ethical affront to non-human animals. Nevertheless, Singer believes that the scale of modern animal agriculture makes it impossible to raise animals under appropriate conditions and to slaughter them humanely. He has, consequently, continued to advocate radical reform of animal agriculture and he has continued to be associated with political groups allied under the banner of animal rights (See Singer, 1979). Peter Singer is therefore one person committed to radical reform at the political level but opposed to animal rights at the philosophical level.

Singer's work has been criticized by philosophers such as Bernard Rollin and Tom Regan, who have found fault in the fact that Singer's reasoning permits abuse of individual animals whenever the compensating benefits for humans or for other animals are great. Rollin and Regan themselves differ, however, on the question of how much reform is called for. Rollin (1981) supports relatively modest reforms of agricultural production systems which would protect an animal's right to satisfy biological needs (Rollin, 1995). Regan (1983), however, argues that animals have a right to live out their natural lifespan, and though he does not stipulate this as an absolute right, Regan nevertheless feels that it is sufficient to require humans to practice vegetarianism under all but the most extreme circumstances. Rollin is therefore an example of a person who might be classified as an animal welfare activist with respect to his political objectives, despite the fact that he is well-known for advocating a philosophical view of animal rights. Regan is clearly committed to animal rights both politically and philosophically.

Conclusion

Although the philosophical distinction between welfare and rights may seem arcane to scientists working on biotechnology, seeing how the welfare/rights distinction is made at different levels of debate is crucial. A call for animal rights is not necessarily inimical to the interests of scientists, nor is a philosopher who, like Peter Singer, adopts an animal welfare rather than an animal rights perspective, necessarily supporting moderate reforms. Selections form the work of Singer, Rollin, and Regan are routinely taught in introductory philosophy classes that stress contemporary moral issues. These writings are attractive to philosophy instructors because they provide a clear contrast of how rights arguments differ from those that stress welfare or utility. A scientist who has obviously failed to master concepts taught in freshman philosophy classes will appear ignorant and unsophisticated to those individuals who received their most systematic exposure to animal issues in such classes.

At this point in time, the relationship between concern for animal well-being and biotechnology is highly tentative. Individuals and groups associated with both animal welfare and animal rights (understood as a political distinction), which are already poised in opposition to scientists, have potential allies for political action in producer and environmental groups and have legitimate questions about the impact of genomic research on animals. To the extent that scientists come to be perceived as lacking compassion or as lacking the ability to address animal issues articulately, the stage is set for confrontation. However, biotechnology can also do much to improve the lot of animals. As such, the confrontational nature of this issue should not be regarded as fixed. Opportunities for communication and better understanding of the issues exist, and it is in the self-interest of scientists in biotechnology to conduct an open and thoughtful review of animal well-being issues (see Thompson, 1997a and b).

9

Ethical Issues and BST

As is evident from its repeated appearance in previous chapters, the controversy over BST involves disputes about many technical issues: Does milk from cattle treated with BST differ from milk now being produced on dairy farms across the nation? How quickly will dairy farmers adopt BST, and how will it effect economics of scale in milk production? Will milk production shift from the traditional dairy states to new locations? How will rural communities be affected? Are we sure that milk produced by BST-treated cows will be properly metabolized by human consumers?

Answers to these and other technical questions are important because they bear upon questions of responsibility, social justice, and human (and animal) well-being. People assume that the answers to these questions entail that BST should or should not be used, but a complete prescription requires additional premises about responsibility, social justice, and human/animal well-being. There are some applications of these concepts on which our society enjoys a firm consensus, but other applications are notoriously contentious. This chapter will examine some of the issues where controversy over ethical ideals and beliefs may lie just beneath the surface of controversy over facts.

Ethics and Unwanted Consequences of Technical Change

One fact of postmodern society is that decisions by a few individuals to develop and disseminate new technologies can have enormous impact upon society as a whole. Although there are many instances where these impacts are predominantly beneficial, there are few (if any) occasions on which they are universally so. Decisions made far from the rural heartland, in corporate offices or in research facilities, can effectively determine that some producers will have to leave farming, that consumers will be buying new food products, and that rural residents, wildlife, and, indeed, society as a whole (including future generations) will have to cope with pollution and resource depletion. The fact that new technologies produce new benefits but may also cause unwanted consequences is the basis for an ethical imperative for decision makers to accept responsibility for weighing benefits against unintended consequences (Jonas 1984).

To some extent, market forces discipline these decisions in an open economy. Consumers may reject products that they think unsafe or otherwise undesirable. The discipline of market forces is limited, however. Sometimes the unwanted effects are not borne by the people with the market power to prevent the technology from being adopted. Competitive pressures sometimes lead producers to make choices that ultimately leave everyone worse off. Lack of information sometimes leads people to make poor choices. Though market forces limit the unwanted consequences of new technology, there may still be important issues of fairness and wisdom in accepting the unwanted consequences allowed by these market forces.

The BST case raises questions about three kinds of unwanted consequences. Papers by Comstock (1988), Burton and McBride (1989), Molnar, Cummins, and Nowak (1990), and by Burkhardt (1991) have summarized ethical issues associated with BST. The first group of impacts are felt by dairy farmers who may be forced to adopt BST (or to cease dairy farming) because of competitive pressures (Browne 1987, Buttel 1986). The second includes consequences for non-human animals (Comstock 1988). The third includes environmental impacts that may bear upon large numbers of people, extending into future generations (Lanyon and Beegle 1989). Food safety issues have less to do with unwanted impact than with uncertainty, discussed below. Milton Hallberg's *Bovine Somatotropin and Emerging Issues* (1992) details the potential for unwanted consequences associated with the introduction of BST and discusses the evidence for concluding whether or not these consequences will occur. Although the range of consequences is quite broad, it is important to see how these three main categories raise different kinds of ethical issues.

1. *Impact upon Dairy Farmers.* There are studies which indicate that BST will increase productivity in herds of those dairy farmers who adopt it. This increase could make it uneconomical for some other dairy farmers to continue in dairying. Although the specific nature of impacts upon dairy farmers and rural communities varies, the broad ethical issue is one of fairness with regard to those farmers who are adversely affected. Will the costs of technical change be fairly distributed?

2. *Impact upon Dairy Cows.* It is the animals themselves that are most

directly affected by BST use. If it is determined that BST harms the dairy cow to which it is administered, this fact will raise the question of whether (and if so, to what extent) the impacts upon animal welfare should militate against use of BST. Do duties to non-humans impose constraints on technology?

3. *Environmental Impacts of Intensification in the Dairy Industry.* If the introduction of BST helps bring about larger dairy farms, as some studies indicate, BST will be implicated in environmental problems (totally apart from the issue of BST) that dairy farmers are facing. If dairy farms that produce much of their own feed and recycle animal wastes are more ecologically sound than dairy farms which must have feed hauled in and waste hauled out, there may be environmental reasons to promote small-scale farms of the sort that may be threatened by BST. Will the adoption of BST exacerbate existing environmental problems?

Each kind of unwanted consequence is ethically controversial. In most sectors of the economy, producers would not expect to be shielded from the economic consequences of technical change. Farmers are raising a concern more typically voiced by organized labor, as when plant closings or new production lines lead to layoffs. Makers of roller skates or stainless steel run the risks of being displaced when new technologies appear in their industries; perhaps farmers should, too. Extension of ethical concern to farm animals and to environmental impacts is also hotly debated.

While Americans share a strong consensus against cruelty to animals, the stresses imposed upon farm animals in food production have not historically been thought inhumane. Given the assumption that animal husbandry practices carried out for the purpose of producing food for humans are generally acceptable, opponents of BST must show why this particular technology is inhumane or, alternatively, must show why traditional standards for animal care should be revised. Environmental quality at one time was thought to conflict with important personal rights of autonomy and control over private property. While there is an emerging consensus that these individual rights must sometimes be overridden by environmental concerns, how public policy is to accomplish this goal remains a matter of considerable debate.

Responsibility and Unwanted Impact

Since there is controversy on each of the three points where BST has been linked to unwanted outcomes, it will be useful to look at two ways of framing the ethical issues of responsibility. In some respects, the fundamental issue at stake in the dispute over BST revolves around which of the two philosophical interpretations of responsibility will best guide our collective understanding of the ethical issues involved. Two points are to be observed in comparing these two approaches.

First, the possibility of radically different philosophical approaches to framing the question of responsibility creates many opportunities for misunderstanding among people and groups who view the issue from alternative vantage points.

Second, each vision of responsibility makes its claims about how a democratic society should go about answering questions raised by unwanted consequences of technical change. As such, this dispute cuts to the core of our political culture and begins to involve issues of far greater lasting significance than BST.

The Intentional-Action Model

Each of three types of unwanted consequence noted above involves impact upon individuals or groups who are powerless to avoid being affected. This is clearly the case with respect to farm animals and unborn generations of human beings, and it is true to a more qualified extent for small-scale dairy farmers, too. Each of us must bear the consequences of events we are powerless to control, as when fire and flood destroy our homes, or when faceless economic forces deplete our savings through inflation.

The creation and implementation of BST is not the result of natural causes or of the faceless machinations of the invisible hand, however. BST is on the scene today because a few hundred individuals made research and development decisions over a half decade. The decisions and the actions that followed were undertaken *intentionally.* No one intended that unwanted consequences would occur; some who participated in research did not even care whether BST ever came on the market. Nonetheless, the unwanted outcomes are reasonably predictable consequences of actions that were undertaken on purpose, and this fact marks a crucial distinction between unwanted consequences associated with BST and those that result from natural or faceless social causes.

The individuals and groups that carried out research and development of BST are capable of actions that impose unwanted consequences on others. The question is whether possession of this capacity gives them an unfair or unjust form of power over others. There are two reasons to think that it might.

1. If those who research and develop technologies like BST have unequal advantages over those who bear the unwanted consequences, there is a *prima facie* reason for raising questions about justice.
2. If the researchers who developed BST are agents who have a charge to protect the interests of those who experience unwanted consequences, there may be a betrayal of trust that raises questions of justice.

Examining BST, we find that the companies developing BST have far more economic power than do small dairy farmers. Researchers developing BST in universities do not face the same threats to livelihood faced by small dairy farmers. Animals and future generations are placed into unequal positions relative to present generation decision makers by simple facts of biology. There seems to be *prima facie* reason for raising questions about justice.

These *prima facie* reasons may be reinforced by the fact that many of the scientists who have participated in the development of BST can be thought of as agents for the general public, at least, and perhaps for the farm community, in particular. A great deal of scientific activity in Western democracies is protected against market forces. Few scientists must turn a profit on their labs. Working at public or nonprofit institutions, many scientists aggressively claim a mandate to do science in the public interest whenever research budgets and academic freedom are at issue. If science is to claim this mandate, it must live up to it. Land-grant universities, where much of the BST work has been done, have historically accepted a further mandate to do science that will strengthen the development of rural communities. As such, dairy farmers may have a special claim upon these institutions. Although no one has argued that scientists have a special responsibility to look out for animals, it is not uncommon or unreasonable to think that the scientific community is especially well-placed to look out for the general public interest in environmental quality. In addition to the scientific community's self-appointed role of public service, then, there are additional reasons to think that two of three areas of unwanted consequence fall particularly in the domain of scientific responsibility.

The fact that BST emerges as a technology for which these considerations are relevant *does not settle* the issue in favor of BST's critics. At most they establish a burden of proof that must be met by anyone who wishes to impose costs upon farm, animal, and environmental interests. Individuals and groups who would benefit from changes may be able to meet this burden of proof, either by argument or by compromise. More narrowly, considerations of fairness and justice establish the right of those who speak for farm, animal, and environmental interests to be heard in the political process. Whether they prevail will depend upon how highly values of equality and trust are rated against needs for food production and economic growth. One might interpret the political debate that has raged over BST as working out an exchange of views in the democratic political process. Whether the handling of BST has been an adequate or effective way for these voices to be heard is a matter that will be taken up below.

The Consequence-Evaluation Model

The idea that any technical change produces winners and losers invites us to think of a new technology as a social bargain in which there are both costs and benefits. Some technologies, as reflected by their capacity to produce an attractive package of benefits at an acceptable social cost, will be a better bargain than others. The key to evaluating this social bargain lies in identifying and measuring the full range of costs and benefits.

A series of value judgments must be made in order to do this. One must establish the value of consequences, such as an illness or a loss of life, that are only indirectly exchanged for other values (such as consumption or economic growth) in the normal course of events. One must decide how to evaluate consequences that occur at some point in the distant future. One must have some way of being reasonably sure that the accounting is complete, and that important costs (or benefits) have not been omitted. While one should not underestimate the difficulty of making these judgments, the idea that a technology's costs and benefits can be compared with the costs and benefits without the technology provides an attractive way of discharging the imperative of responsibility for technical change.

When applied to BST, the consequence-evaluation model regards adverse, unwanted outcomes as costs weighed against the projected benefits derived from lower milk costs. It is possible that there could be environmental benefits, as well as environmental costs, thus the framework for such an evaluation cannot be set in advance of empirical investigation. While a reduction in milk costs might be slight for an individual consumer, they may be quite large when multiplied by the large number of people who use milk. As such, it is quite possible that the benefits of BST may outweigh the costs. The actual assessment of costs and benefits requires a significant amount of technical expertise. When one adopts a consequence-evaluation model for assessing new technologies, the question of whether BST is an ethically acceptable technology hangs upon the answers to these technical questions.

Notice that when one compares total outcomes from two or more options (at a minimum, the options include BST and no BST) there is no obvious reason why intentional action should enter the picture at all. There are costs and benefits associated with the *status quo,* even when they are the result of faceless economic forces that have evolved through time. If the trade-offs between cost and benefit for the *status quo* are clearly less attractive than the trade-offs from technical change, the consequence evaluation model suggests no reason to place additional significance upon consequences of international actions. If an analyst is convinced that equality or autonomy *is* important, these factors enter the consequence-evaluation model as additional sources of value that must be measured if the accounting is to be complete. Assessing the consequence value of a loss of autonomy may prove quite difficult, and it is, in any case, a very different view of how autonomy figures in the imperative of responsibility than that developed in the intentional action model. The possibility of taking very different approaches to the problem of unwanted outcomes can itself feed policy controversy. It is far easier for two people who have different interpretations of responsibility to talk past each other than it is for them to communicate. Even more fundamentally, some who reject the consequence-evaluation model see it as contrary to traditional values, or as a technocratic response to political problems that require open-ended political dialogue (Sagoff 1988).

These issues are taken up again in the later sections of this chapter. First it is important to survey the ethical importance of consumer concerns that have dominated much of the public discussion of BST.

Ethics and Uncertainty

By its very nature, technical change involves unprecedented events. Although it is possible to predict certain kinds of consequence with reasonable confidence, these predictions are themselves seldom uncontested. The reality of disagreement among alleged experts creates a situation in which a member of the lay public, lacking even the evidence to make informed judgments about who to believe, quite reasonably comes to regard all claims about the likely consequences of technical change with justifiable skepticism. Faced with a lineup of Ph.D.s expressing contradictory claims, the educated layperson has no alternative but to bring highly subjective factors to bear in deciding who to believe. Since it is logically impossible for all the experts who espouse contradictory views to be speaking the whole truth, it is very reasonable to question the validity of claims that any expert makes about the true risk of a technology. This skepticism is amplified when alleged experts betray their insensitivity to a layperson's dilemma by making statements about the ignorance, irrationality, and emotionalism of public concern, or by describing these concerns as perceived rather than real. It is quite rational to doubt the judgment of a person who seeks your support by calling you a fool.

The unfortunate upshot is that political decisions about technology often become dominated by uncertainty. Technical uncertainty creates an opportunity for experts to disagree. When experts disagree, non-experts are faced with uncertainty about who to believe. When experts become advocates for a technology, non-experts have reason to suspect that their advocacy is based upon personal interests. Scientists who criticize public judgment may be attempting to establish positions of power from which they can advance their research interests, their prestige, and ultimately their financial gain. In such a political environment a lay person may reasonably conclude that information about who has the most to lose or gain by adoption or rejection of technical change is highly relevant to decisions about who to believe. (Thompson 1986)

The ethical character of the BST debate changed drastically when claims about the safety of drinking milk from cows treated with BST became a contested issue. Until then, the general issue was how to resolve questions of responsibility for the unwanted consequences of introducing BST. These consequences were of political importance to those who spoke for small farms and animal and environmental interests but might have been overlooked by many who take an active interest in public affairs. With the advent of controversy over food safety, the potential spectrum of affected parties increased dramatically.

What is even more important is the way that the ethical issue shifted from being one of dealing with unwanted consequences to one of uncertainty. This is a subtle point, but one that should not be missed by students of the BST case. There has been no serious scientific evidence to suggest unwanted health consequences for consumers of BST milk. Consumer groups reacting to the food safety issue were not reacting to a health risk *per se,* even a negligible one, as in the case of Alar on apples. Consumer groups were reacting to uncertainty, to a problem in deciding who to believe about BST and milk. While the fact that an overwhelming majority of credentialed scientists see no health risk associated with BST milk should (and, in the long run, probably will) count heavily in favor of BST, the fact that many of these scientists are linked with private corporations and research institutions that stand to gain from sales of BST weighs in against it.

Another subtle distinction should not be missed: the layperson does not evaluate the risk of BST as such. The layperson must evaluate the risk of choosing the wrong expert. Arrogant behavior and an inability to use common English are relevant pieces of evidence for the decision, "Who should I believe?" In the BST case, consumers may rationally see little cost in being deceived by BST's opponents and may be comparing this risk to a low probability/high consequence risk of being deceived by the majority of voices speaking for the FDA and the scientific community.

In understanding the way that ethics bear upon risk and uncertainty, it is crucial to see that all of the consumer's information about the safety of BST is subject to a conditional probability that the source of that information is either ignorant or, worse, dishonest. Since a component of the layperson's risk estimate derives from uncertainty over sources of information, the risk will not be calculated directly from scientific studies. Many authors have taken to describing the difference between risks calculated on the basis of scientific evidence and risks calculated on the basis of corrigibility of human beings who report scientific findings as a distinction between "real" and "perceived" risk (Rowe 1977, Fischhoff and co-authors 1981, Rescher 1983, Johnson and Covello 1987, Lewis 1990, Glickman and Gough 1990). This choice of words is sometimes unfortunate, for it can be taken to imply that the lay person is responding to extraneous and irrelevant evidence. While it is almost certainly true that many of us do make erroneous and even irrational risk judgments, evidence bearing upon an information source's willingness and ability to report the truth is hardly extraneous or irrelevant to a layperson's estimate of food safety risk.

Given the background of the uncertainty problem faced by food consumers and consumer advocates, it is not surprising that the issue evolved into a debate about the risks of BST and milk. The scientific community has come to view risk issues as an expected value problem, and this is the way it has approached the food safety issue for BST. While there are clearly many cases in which the assessment of expected values is the right approach to take for food safety, alternative burden-of-proof approaches may have been a better choice for BST.

Alternative Ethical Approaches to Risk

As can be seen from the discussion of uncertainty, risk problems are extremely diverse and complicated. Two general approaches to risk issues can be outlined, though there are important variations and varieties of each.

1. *Expected-value approaches.* These approaches assume that risk is defined by at least two key variables: the probability and value of events. If these variables are determined, it is possible to interpret risk as an expected value, i.e., as a set of outcomes anticipated with the given degree of probability.

2. *Burden-of-proof approaches.* Burden-of-proof approaches associate risk with actions that will ultimately be sanctioned, prohibited, or allowed with qualifications or modifications. Sanction depends upon meeting burdens of proof that may or may not require one to assess expected values. If the individuals who voluntarily consent to participate in an activity also bear the risks of the activity, it may be possible to permit an action with little or no assessment of expected values.

The long-standing practices of the Food and Drug Administration (FDA) have established a tradition of taking expected value approaches to the problem of food-borne risks. A vast battery of toxicological tests can be brought to bear in assessing probable harm from using a food substance or additive. The expected-value approach may have seemed especially applicable to BST, for researchers had reason to think that measured risks for BST milk would be identical to those for non-BST milk. This judgment would have been correct if the issue had been one of unwanted consequences, or if risk assessment been performed in a political environment of attempting to find evidence for unwanted consequences. Having found none, one might have presumed that the food safety issue was at an end.

Within the context of an uncertainty problem, however, expected value approaches to risk can backfire, providing the layperson with more reason to doubt expert opinion. The reason is simple. From the layperson's perspective, risk assessors are just one more group of self-professed experts. The criterion of assessing the effects of a particular decision on their interests should be applied. If FDA and university scientists are judged to have close working relationships with the companies and scientists who have promoted BST, publication of assuring risk estimates may just as well be interpreted as a ploy, as objective, unbiased reporting of evidence of safety. Responding to uncertainty problems with technical risk assessments is, to a person unschooled in probability and consequence evaluation, little more than saying, "Trust me."

Uncertainty issues are politically fractious and intense. It is far from clear that burden-of-proof approaches would have fared better. It is possible, however, that an agreement to label BST milk might have been interpreted as a gesture of good faith. Those who choose to accept the dominant judgment of the scientific community could capture price savings on milk. Those who chose to reject that judgment for whatever reason are free to do so. The label alters the burden of proof that must be offered for BST milk from one of having to trust potentially self-interested parties to one of being in a position of choice.

Companies involved in the manufacture of BST have wished to avoid labels because they fear that labels imply a stigma that will hurt sales. FDA officials have resisted the strategy because they fear that such labels might imply a health benefit where none is known. As such, labels are hardly a panacea. While labels satisfy a burden of proof for acceptable risk, requiring labels may have policy implications that are themselves unacceptable.

The emergence of consumer uncertainty over the safety of BST has blown the political issue out of proportion to its significance as a matter of unwanted impacts. While the impacts of BST are important issues, especially if environmental concerns prove to be warranted, they do not pose a threat to democratic institutions for public decision making in the way that uncertainty issues do. With respect to our long-term prospects for governance of science and technology, BST is probably a minor episode, but we should not overlook the opportunity to learn from it. Some of these larger philosophical issues are taken up in the next, and final, section.

Ethics and Political Consensus

Democratic political theory has evolved around the concept of a social contract. Government arises because free and informed individuals have agreed to be bound by a sovereign political of authority. Such agreement is obviously an idealization, perhaps even a metaphor, and the social contract idea has had its share of detractors throughout history. Nevertheless, contract ideals have based the authority of government and public institutions squarely upon consent of the governed. They have consistently opposed metaphysical claims to authority, regardless of source-divine right, textual hermeneutics, or historical materialism. The norm of contractual consent is based upon a recognition that people must find a way to live together, and that stable social expectations about how power will be exercised and how disputes will be resolved are in everyone's interest.

There was a time when the philosophy of scientific inquiry was deeply allied with philosophical arguments for democratic political structures. The ideal of reproducibility was thought to make science an inherently democratic social institution. Under the leadership of

Robert Boyle, the British Royal Society sought to make public demonstrability of experimental results the *sine qua non* of scientific inquiry. In Boyle's view, such a standard assured that scientific truth would rest upon propositions whose validity was apparent to everyone and would create a situation which public acceptance of validity did not depend upon the scientist's position in political, religious, or academic hierarchies (Shapin and Schaffer 1985). Although he fought bitter battles with Thomas Hobbes, the originator of social contract theory, Boyle's belief that scientific truth was essentially founded upon public agreement about the incontestability of facts has been imitated by political theorists from Locke to Dewey and Rawls. These social contract theorists argue that democratic government must emulate scientists' practice of arriving at truth by public debate within a community of inquiry.

BST has tested the social contract. Researchers and private companies have undertaken research and development on BST with the expectation that, if the product finds market acceptance, their efforts will be rewarded. While it is reasonable that they should have expected to deal with some of the unwanted consequences of BST, it was not reasonable to expect that food safety issues would be among them. The emergence of uncertainty and, in turn, the food safety issue is evidence of trouble in the contract. It is evidence of a lack of confidence in science and in its institutions. This is a development that should be viewed as quite serious, not only for science, but for the foundations of democratic institutions.

The problem is that both commerce and political decision making require a certain amount of trust. Trust must, of course, be won and once won must be preserved. The matter of why the American public is now so skeptical of science institutions and the biotechnology industry is the subject of a much longer discussion than can be attempted here. Whatever the causes and however just or unjust the suspicion of science might be, the largest and most serious ethical issue associated with BST is the matter of trust. All the other ethical questions feed into this one. The matter of unwanted consequences becomes an ethical issue because we have learned how decisions to research and develop technology can lead to costs that are not borne by the people who make the decision or who reap the benefits. It is impossible to avoid all unwanted consequences; stifling technical change has unwanted consequences, too. The ethical problem is that we must trust people who do not bear the costs to look out for the interests of those who do. The issue of uncertainty is even more transparently related to the issue of trust. Even well-educated citizens are placed in a position of needing to trust the statements of scientists on technical issues of risk and safety, but the multiplicity of voices speaking on an issue makes this trust hard to come by. Repressing dissent is hardly the solution, for such a strategy not only violates the basic political liberty of free speech but is likely to increase public suspicion of scientists' motives.

The idea of a social contract is used to help us understand the range of authority that should be vested in government. The social contract can be seen as a bargain struck among all members of society in which the unrestricted freedom that would exist in a lawless state of nature is given up in exchange for the state's protection of more basic rights, such as life, liberty, and property. It is only recently that scientific research has begun to play an important role in this contract.

There are two reasons for the change. First, science is now seen as essential to government's ability to protect the life and health of its people. Science is instrumental in identifying threats to life in the form of disease or trauma risks that would otherwise appear to be "acts of God." Second, science's role in developing technologies having unwanted consequences means that science can itself be seen as a threat to a person's livelihood, and perhaps even to life. Science is thus put in the position of being both a threat and a guarantor against threats. To some, this may appear to be a case of asking the fox to guard the hen house.

One way of solving this problem is to build a high wall between that component of science which is in a position of public trust and that portion of science which is involved in the development of technologies that may produce unwanted consequences. Public science, conducted at non-profit institutions, would enjoy public confidence. Private science, conducted in the private sector, would be held to the same degree of accountability normally expected of any commercial activity. The flaws in this solution are complicated and subtle. Although they cannot be examined in detail here, it is worth noting four features of contemporary science that make this ideal very difficult to achieve.

1. *Scientific research does not respect the public/private divide.* Often the same basic science underlies regulatory science and technology development alike. Scientists working on key theoretical topics are likely to be colleagues, without regard to their public or private employers, and scientific knowledge is likely to flow back and forth across this divide.

2. *Enforcing a strong separation between public and private science is impractical.* Individual scientists are likely to go back and forth between public and private research, particularly when technologies are closely linked to important theoretical developments, as was the case with BST. Private scientists will have been educated in non-profit institutions and will naturally maintain friendships there. Any attempt to control such interaction would be an unacceptable intrusion upon the private life of scientists, and enforcement would be enormously expensive.

3. *Public science institutions are finding it necessary to cultivate private sources of research funding.* The costs of scientific research have escalated beyond the capacity of public and foundation resources. Public researchers have adopted strategies such as contract research, joint public/private positions, and the seeking of patent protection for their discoveries as a way to assure that the dollars needed for future research will be secure. Without a massive increase in government expenditure for research, public researchers will continue to depend upon private money for the foreseeable future.

4. *A strong separation between public and private science sectors might well weaken public science.* The three preceding problems create a situation in which good scientists might well flee public institutions if they were forced to sever contacts with the private sector. For the time being, a post as university professor on a United States campus remains attractive to many of the world's best scientists.

The dilemma, therefore, is deep. The tension between the regulatory and the technology-stimulating roles of science erodes public trust in science institutions. At the same time, any solution to this problem must be sensitive to the delicate network of personal relationships that make science possible.

Conclusion

The ethical controversy over BST arose because, like many technologies, it may produce some effects that are unwanted. There is no reason to think that the unwanted consequences of BST are particularly dramatic or extreme, but the fact that decision makers within public research organizations and private companies can affect others makes these unwanted outcomes an issue of some significance. The significance has escalated, however, because of the food safety questions that have been raised, and because of the climate of uncertainty that they generated. It is the uncertainty issue that truly threatens to keep BST off the market at this writing,[1] and it is one that the developers of the technology had no reason to expect.

This, in turn, leads to the questions of trust that are crucial to democratic institutions. This is not to say that the success or failure of U.S. constitutional democracy hangs upon the BST decision, but this policy problem can be expected to recur in the future with respect to other technology. American society must resolve whether we can expect to develop biotechnology products in an orderly and efficient manner.

1. FDA approval of BST was still pending when this chapter was written but was granted in 1993. The U.S. Congress immediately enacted a one-year moratorium on the use of BST, which expired in 1994. However, even as this volume goes to press in 1998, BST has not been approved for use in Canada or in most of Europe. See Thompson (1997a) for more details.

10

Animals in the Agrarian Ideal

The chapter begins with the premise that because most Americans live in cities, their image of animal agriculture has been formed by books, pictures, stories and by portrayals in newspapers and on television. From this premise, the chapter will argue that urban Americans' understanding of animal agriculture is based upon a philosophy of agriculture that might be called "the agrarian ideal." I shall examine a set of concepts drawn from writers who attempted to form a unified and action-guiding philosophy of agriculture during the eighteenth and nineteenth centuries, particularly Thomas Jefferson and Ralph Waldo Emerson. After completing my sketch of the agrarian ideal, I will draw upon a more diverse and popular literature to illustrate how animals fit into this ideal.

Two important empirical questions limit my approach. One is whether the agrarian ideal does, in fact, influence anyone's thinking on animal agriculture. The popular literature discussed below concentrates on farm animals, but urban Americans are exposed to many other portrayals of animals. These range from the totally anthropomorphic Teenage Mutant Ninja Turtles to the naturalistic animal films on *Wild Kingdom* or *National Geographic*. There is thus an open question as to how much influence these images have upon someone's understanding of agricultural animals. Assuming there is some influence, the second question is how the agrarian ideal is transmitted, how a nation of people far removed from farms and from the agrarian literature of the nineteenth century could, nevertheless, come to be influenced by agrarian ideals. I have assumed that the influence comes both from explicit statements of the ideal and from literary and artistic images. Empirical demonstration of these assumptions lies beyond the scope of this book, as well as my competence as a researcher.

My discussion of the agrarian ideal is not intended to suggest that most people have emotional, romantic, or uninformed opinions of animal agriculture. My goal is to examine how the agrarian ideal can be thought to entail norms for the use of animals in agricultural production. To do this, I will show how the agrarian ideal can be contrasted with a utilitarian philosophy, and how agrarian philosophy suggests an alternative framework of philosophical values for understanding animal production. There can be no doubt that the agrarian ideal is far from the reality of contemporary agriculture but, while that distance might be taken to show that the agrarian ideal is misinformed, it may also be taken to show where contemporary agriculture has lost its way. I shall return to this possibility only at the very conclusion of my remarks.

Two Philosophies of Agriculture

The agrarian ideal is a philosophically based vision of what agriculture would be under ideal conditions. It therefore presents a target, goal or standard for actual agricultural practice in that deviation from the ideal can be regarded as an imperfection, and in that changes which are judged to increase divergence between reality and the ideal are understood as degenerate and regressive. In contrast to positing a norm as an ideal, norms function more commonly as rules which prescribe action of one sort or another. The *utilitarian maxim,* to act so as to produce the greatest good for the greatest number of people, is an example of such a norm. The utilitarian maxim has often been applied to choices of agricultural production techniques. Among the most straightforward and elegant applications is John Locke's defense of enclosing the commons for private production, cited in the introduction, but worth repeating here.

> He that encloses land, and has a greater plenty of the conveniences of life from ten acres, than he could have from an hundred left to nature, may truly be said to give ninety acres to mankind: for his labour now supplies him with provisions out of ten acres, which were but the product of an hundred lying in common. (Locke, 1989 reprint, p. 23-4.)

The persistent application of an injunction to "make two blades of grass grow where one grew before," is what I shall refer to as the utilitarian philosophy of agriculture. It is an ethical principle that evaluates agricultural practices in light of their ability to produce increases in yield or quality of agricultural products, though it can also be readily adjusted so that beneficial outcomes are weighed against costs, both to producers and to the general public (Thompson, Ellis and Stout, 1991).

Chapter 1 offered a discussion of the utilitarian maxim and the agricultural research that guided the development of husbandry practices during the past century. Rosenburg (1976) and Danbom (1986) have also contributed excellent historical studies of the utilitarian ideal in agricultural research. It is the profitability of farming that has been the immediate focus of both producers and researchers, but Chapter 1 shows how the goal of assuring or increasing profitability can itself be ethically justified in utilitarian terms by showing how new agricultural technologies assure food availability and decrease costs for consumers. This is an argument which has been mastered by agricultural researchers, and it need not be repeated here. It is important to see that the utilitarian maxim provides the basis for a philosophy of agriculture, and that it expresses a criterion for making value judgments about the acceptability or justification of particular production practices. It is also important to see that agricultural research (as well as husbandry practices), which stresses productivity, is not morally neutral. Productivity-enhancing animal science is an expression and application of a utilitarian philosophy of agriculture.

The agrarian ideal provides an alternative criterion for evaluating a system of agricultural practices. It is a view of agriculture that sees goodness in small, family-owned and -operated production units (or "farms," as an agrarian would call them) and typically assumes also that production activities are highly diversified. It is, in caricature, the family farm on which Pa raises wheat, corn, and potatoes, while Ma tends the chickens, ducks, and geese. The livestock include cattle and hogs. They seem to pretty much take care of themselves, though they may require daily feeding. The children have milking and gardening chores, while Grandma and Grandpa occupy themselves with quilting, making jams and jellies, repair of tools and other work which maintains the household. A philosophically respectable account of this ideal is discussed below, but it is worth sketching in caricature just to emphasize how it differs from the utilitarian philosophy.

An agrarian philosophy will specify certain roles and activities that occur on healthy or prototypical farms, and while it is clear that few (if any) farms fit the ideal exactly, a given farm, as well as the structure of agriculture generally, can be compared to this picture to provide an evaluation of where reality stands with respect to it as a standard. It is doubtful that the agrarian ideal will be action-guiding in the same way that the utilitarian maxim is. Each farm will be different, and indeed, the agrarian ideal celebrates the uniqueness of farms. Nevertheless, it provides an image of how farming *in general* should be, and this image can be used as a standard to judge the pattern of farming practices at any given time. Agrarian and utilitarian philosophies are not logically incompatible. It is possible that a change which moves agriculture closer to the agrarian ideal would also produce the greatest good for the greatest number. While the two philosophies might recommend the same plan of action under certain circumstances, they go about justifying such a recommendation in very different ways. Furthermore, it is likely that under contemporary conditions, they are unlikely to converge with respect to agricultural practices.

Agrarianism

A full fledged agrarian vision of agriculture requires fairly lengthy development. Wendell Berry (discussed in Chapter 5) is the best contemporary expositor of agrarianism. Others include Harold Breimyer (1965), Jim Hightower (1975a and b), Frances Moore Lappe (1985) and Richard Cartwright Austin (1990). These authors present agrarianism as a critique of what has gone wrong in agriculture. The agrarian ideal is described, and contemporary agriculture (which fares well under utilitarian criteria) is found lacking. The agrarian ideal can also be presented as a demonstration of what was right about the agriculture of eighteenth- and nineteenth-century America. Everyone's favorite agrarian is Thomas Jefferson and it is worth elaborating the exposition given in Chapter 5. In his book, *Notes on the State of Virginia,* Jefferson wrote:

> Those who labour in the earth are the chosen people of God, if ever he had a chosen people, whose breasts he has made his peculiar deposit for substantial and genuine virtue. It is the focus in which he keeps alive that sacred fire, which otherwise might escape from the face of the earth. Corruption of morals in the mass of cultivators is a phenomenon of which no age nor nation has furnished an example. It is the mark set on those, who not looking up to heaven, to their own soil and industry, as does the husbandman, for their subsistence, depend for it on the casualties and caprice of customers. Dependence begets subservience and venality, suffocates the germ of virtue, and prepares fit tools for the designs of ambition. This, the natural process and consequence of the arts, has sometimes perhaps been retarded by accidental circumstances: but, generally speaking, the proportion which the aggregate of the other classes of citizens bears in any state to that of its husbandmen, is the proportion

> of its unsound to its healthy parts, and is a good-enough barometer whereby to measure its degree of corruption. While we have land to labour then, let us never wish to see our citizens occupied at a work-bench, or twirling a distaff. Carpenters, masons, smiths, are wanting in husbandry: but, for the general operations of manufacture, let our work-shops remain in Europe. It is better to carry provisions and materials to workmen there, than bring them to the provisions and materials, and with them their manners and principles. The loss by the transportation of commodities across the Atlantic will be made up in happiness and permanence of government. The mobs of great cities add just so much to the support of pure government, as sores do to the strength of the human body. It is the manners and spirit of a people which preserve a republic in vigour. A degeneracy in these is a canker which soon eats to the heart of its laws and constitution. (Jefferson, 1984b, 290-291.)

This passage is often cited by those who wish to praise farming, but its true meaning can only be grasped by recognizing the context in which it originally appeared.

Jefferson's attention was constantly focused on political and constitutional issues, and his *Notes on the State of Virginia* was widely read as a reference work for the constitutional debates of 1783. A key point of debate was between Federalists, such as Madison, Hamilton, or Jay, and Democratic-Republicans, such as Jefferson. Federalists feared democracy on philosophical grounds that stretch back to the Socratic dialogues. The problem with democracy lay in the threat of mob rule, of short-sighted and freeriding masses who would insist upon receiving benefits from government, but who would not accept responsibility to secure the foundations of the state. Democrats, it was feared, would accumulate public debt to secure benefits and to pursue adventures but would not invest in the institutions—education and infrastructure—needed to secure a sound future.

This is an argument that had been used to defend the aristocracy in Europe. Landowning gentlemen would have personal interests that depended upon a strong state, hence they would see no conflict between their role as citizens and their livelihood. Merchants and tradesmen, by contrast, could vote themselves benefits and abandon the state when the bills came due. Ownership of land tied aristocratic wealth to an immovable asset. This fact marked the key differences for determining who would (and who would not) be a good citizen. Jefferson saw that what applied to aristocrats in Europe applied to a majority of the population in the new United States. Farmers are valuable *as citizens* because they identify personal and public interest, and because their ties to the land preclude them from supporting irresponsible courses of action. Jefferson is saying rather little about farmer's broad moral virtues, but he is arguing that *citizenship* is a virtue which comes naturally to the farmer because of the farmer's dependence upon land, rather than skill or capital as an economic asset.

The notion that farming promotes citizenship is the bedrock of the agrarian ideal. The ideal was expanded to include a broader set of moral ideas when it was linked to a series of character traits that were deemed both spiritually and politically praiseworthy. Not the least of these was a concept of productivity that ties citizenship and moral virtue to industriousness. Farmers who were successful worked hard. Their reward for hard work was in their agricultural bounty. Their industriousness made them valuable as citizens because their hard work made them appreciate the fragility of the public good all the more. By investing so much of themselves in their farms they came to identify with land and tended to place even greater stock in its continued thriving.

In addition to citizenship and industriousness, a third value in the agrarian ideal is self-reliance. As developed by Ralph Waldo Emerson, self-reliance does not imply material independence from others (as, for example, does our contemporary idea of food security). Instead, Emerson meant the development of a psychological capacity to accept moral responsibility for one's fate. Ironically, this means that self-reliance stresses the interconnectedness of things. To succeed as a nineteenth-century farmer, one needed to understand how climate, soil and production were interdependent. Farmers diversified to produce the full range of needed commodities, but also because doing so reduced the potential for catastrophic loss. More importantly, one needed to understand how the roles of the household were interdependent. Children of nineteenth-century farm families learned to do their chores because they could see that failing to do them had consequences for the entire family. Each member of the family had role-defined tasks that were essential for family well-being.

These three values—citizenship, industriousness, and self-reliance—do not exhaust the philosophical foundations of the agrarian ideal, but they must suffice for the present account. The point of the ideal is that particular ways of life would naturally lead people to develop these virtues. The virtues are both intrinsically valuable to the individual who possesses them and functionally valuable for democratic society and for stewardship of natural resources. Farmers develop these virtues because farming is a way of life that unifies the moral values of these virtues with the practical imperative of self-interest. At least a certain *kind* of farming will do this. The kind of farming that is required is clearly one in which the landowner works the land personally and with the intention of bequeathing the land intact to descendants. It is a kind of farming that is diversified and which has identifiable roles for each member of the family. It is family farming, in the classical sense.

Animals in the Agrarian Ideal

Animals figure importantly, but not prominently, in the agrarian ideal. Nineteenth-century authors mention animals as a matter of course, but do not provide extended discussions of animals. Boyd Smith's *The Farm Book,* first published in 1910, was intended to educate city children about farming. It provides one source for examining the role of animals in the agrarian ideal. Family farms are expected to have livestock and poultry, as well as work animals such as oxen, horses, cats and dogs. These animals are part of the diversity upon which the family depends. Family dependence upon work animals gives them a special status: they are not commonly used for food on U.S. farms, even when farm existence was hardscrabble and when the occasional meal of dog or horse meat would have made an excellent source of protein. Horse meat *was* eaten, of course, but not commonly. Whether kept for food or work, however, animals had their place as part of the support network for the family. They were an integral part of the farming operation, but one does not find prominent essays on the role and function of animals in the agrarian ideal until comparatively recent times.

There is, for example, a fairly extensive discussion of draft animals and horse farming in Wendell Berry's 1977 book, *The Unsettling of America.* Berry begins by analyzing arguments (attributed to economist Earle Gavett and to then Secretary of Agriculture Earl Butz) that are intended to show how many draft animals it would have taken to substitute for the number of tractors in use in 1975. Gavett's figure of 61,000,000 horses and mules was 58,000,000 million in excess of the animals then available. Berry takes issue with this projection, apparently intending to show that an agriculture based on draft animals is a meaningful alternative for our own time. The details of his argument are not relevant to the issue at hand, but it is worth quoting part of his critique of Butz:

> It is easy to say, as former Secretary Butz said in his own fatuous attack on the "anti-technological revolution," that "To return to the 'good old days' in agriculture, or indeed just to cling stubbornly to the farm-

> ing methods of today, would be to condemn hundreds of millions of people to a lingering death by malnutrition and starvation in the years ahead." But that is simply the oldest—and the most profitable—cliché of the industrial revolution, supported only by a thoughtless obeisance to "progress." (Berry, 1977, p. 204.)

Butz's rationale is utilitarian: we evaluate agriculture by examining the consequences in terms of cost and benefit. Berry rejects the utilitarian philosophy out of hand. Berry follows this attack with a description of three horse-powered Iowa farms, from which he concludes:

> It will be observed that the use of horses is not just a means of doing work, a kind of power added to a farm from outside as petroleum or electricity is added to it. The use of horses is a means that *belongs* to the farm; it is a way of farming; it is, as Maurice Telleen points out, invariably accompanied or followed by a set of practices that belong together. If made to belong to the land by good care and good sense, horses tend to preserve its health. With horses come pastures and hay fields, because the horses must eat. And if one is going to grow forage for horses, then one finds it natural and economical to grow it also for other animals. From the growing of forage and the diversification of animal species, there follow naturally the principles of diversification and rotation of field crops. Having animals, one has manure, and so manure is used instead of commercial chemical fertilizers. And the use of manure, the conservation of humus, and the practices of rotation and diversification tend to work against diseases, insects, and weeds, and so one uses few or no pesticides. It is a way of farming that involves year-round use of the land by animals, plants, and the farm people—in contrast to the "corn, beans, and Florida" rotation of orthodox cash-grain farmers. Moreover, the farmer who farms with horses is not likely to be an expander. His way of farming tends to confine him to a limited acreage near home. He therefore concentrates his attention and, instead of getting more, takes good care of what he has—sows cover crops, guards against erosion, etc. (Berry, 1977, 208-9.)

This paragraph, as well as Berry's discussion of Amish farming practices in *The Unsettling of America,* places draft animals into the constellation of practices that were thought to promote moral character in the agrarian ideal. The use of horses promotes diverse farming practices, which, in turn, promote stewardship. The horse farmer works hard, but well. Because the horse farmer is confined to "acreage near home," he is more likely to see the farm as exhibiting an organic whole, as providing social roles and categories of self-knowledge for achieving an integrated interpretation of the full range of life activities. Although yields on the horse farm may be relatively small they are sustainable, promoting self-reliance. In short, Berry makes farming with horses into a component of the agrarian ideal. He shows how farming with draft animals locks the farm family into a pattern of activities commonly associated with the agrarian vision of what good farming is. By contrast, mechanized agriculture frees farmers from bonds in ways that promote, not self-realization, but profligacy, specialization, and, ultimately, brittle and unsound farming practices.

Animals and the Ideal of Innocence

Animals *do* have a cultural significance for nineteenth-century writers that should not be overlooked, however, and it differs in important respects from Wendell Berry's. One of Emerson's themes is how city life—the life of the mind—imposes unnatural rhythm and patterns upon the human spirit. For Victorians, childhood was a time of innocence, a time in which human nature shone forth; but maturity consisted of the bridling of natural passion, passion which became disruptive and destabilizing for adults. It is childhood that is the important Victorian concept, but note how Emerson links the idea of innocence to both farming and to animals. In his essay *Farming* he writes,

> That uncorrupted behavior which we admire in animals and in young children belongs to him, to the hunter, the sailor,—the man who lives in the presence of Nature. Cities force growth, and make men talkative and entertaining, but they make them artificial. What possesses interest for us is the *natural* of each, his constitutional excellence. This is for ever a surprise, engaging and lovely; we cannot be satiated with knowing it, and about it; and it is this which the conversation with Nature cherishes and guards. (Emerson, 1904, p. 153-154.)

The obscurity of this passage should not be allowed to overshadow its importance. When Emerson talks about constitutional excellence, he is using the word "constitution" in a sense closely akin to the way that we talk about a person's constitution when we refer to their general state of health, as in "He has a sturdy constitution." The word also connotes basic or foundational traits of character, traits that are molded by one's nature and by one's life experience. In this passage,

Emerson hints at the agrarian way of understanding how nature and experience are linked. A person has a nature, a *natural,* which establishes one's potential for virtue and for excellence. One attains excellence when one's experience is fitted to one's nature in such a way as to encourage the development of that potential. Morality for Emerson consists not in the following of rules or the optimizing of trade-offs but in the realization of self: true self-reliance. Just as a person attains excellence by living up to the implicit potential of her constitution, so does a society. This theme, drawn here from the writings of Emerson, can also be traced to Rousseau.

Farmers and animals are important because they follow the call of nature. The lives they lead are more truly guided by their implicit *natural* (to repeat Emerson's term) than are lives lead in cities, where fashion, cleverness and imitation call the tune. Farmers live a life, the agrarian life, which promises realization of constitutional excellence. The tasks they must perform and the roles they must play are all set by nature, by the farmer's need to accommodate his capacities to nature's demands (and simultaneously to satisfy his own demands in terms of nature's capacities). The Jeffersonian message was that, just as a person's constitution is fulfilled by the agrarian life, so is a society when its constitution is built upon a nation of farmers. Combining Emerson with Jefferson, then, both personal and social virtues (the character traits needed for citizenship) depend upon a person living a life that expresses and develops evolved human capacities. The small farm is an environment in which this is likely to happen.

Animals also become important as object lessons. Emerson well knew that many people would not live the life of the farmer. Indeed, he preferred the life of the poet for himself. Poets, preachers, tradespeople and city dwellers may not literally live the life of the yeoman farmer, but they will be well-served if they contemplate the life that the farmer lives. The city is a destructive environment because the human tendency to imitate is no longer guided by nature. It takes off on an errand of pure artifice, resulting in a condition in which people live lives that have no natural foundations at all. As such, it becomes part of one's duty to at least contemplate the natural and to make every effort to maintain a kind of innocence. For Emerson, this was possible because one lived in community with natural beings who formed a kind of hierarchy of innocence. Children were innocent not because they were uncorrupted by lust or ambition, but simply because in their undeveloped constitution, nature weighs far more heavily than experience. Farmers, fishers and *animals* live lives in which nature and experience are fitted like hand and glove. They have a kind of innocence which is admirable, which all should strive to attain. Animals thus make good role models for people. They have a natural way of living that should inspire poets, preachers, and ordinary mortals.

The Agrarian Ideal Today

The literary statements of the agrarian ideal, from Ralph Waldo Emerson to Wendell Berry, reinforce two lessons with respect to animals. The first is that animals have a role in the integrated and diversified farm that is the agrarian model of a healthy human environment. These animals provide for their human masters, but humans must also care for them. Learning to care for animals, and learning to recognize one's dependence upon them, is part of the character-forming experience that makes farm life so valuable. The second lesson is that, in a limited sense, animals can serve as role models for human beings. Animals on nineteenth-century farms were thought to live lives for which they were constitutionally suited. Since this is what we all should do, there is something to be admired in the ease with which animals do this. One can say that these two lessons establish a framework in which we can determine the value of animals for an agrarian. It is not the sort of value that can be exchanged for a substitute. It is value that emerges and informs the human character *constitutionally,* by establishing the framework in which other values are to be exchanged.

There are both philosophical and empirical questions that might be raised about the agrarian ideal. It may, for example, rest upon naive or even maliciously false ideas about human and animal nature, about constitution and character, or, indeed, about the nature of ethics and morality. Some of these questions point toward intellectual history. The Victorian age was, in some respects, an intellectual contest between visionaries such as Emerson, who understood innocence as something to be achieved by aiming at one's "*natural,*" and reactionaries who assumed innocence could only be lost through surrender to natural impulses (meaning sex, mostly). A different set of questions are raised when we ask how the agrarian ideal continues to be transmitted in American culture. How can agrarianism possibly have any influence in a society where most people see farms only from behind a windshield (if that), and how are agrarian ideals themselves affected by an urban life experience in which one's typical contact with live animals is confined to pets, zoos and vermin? The researchable topics raised by these questions lie beyond the scope of my book. However, we can obtain some insight into the purely philosophical aspects by looking at children's books.

The Store-Bought Doll (Meyer, 1983) is a Golden Book set around the turn of the century, but published in 1983.[1] The book is a simple story consisting of three parts. In the first part of the book, Christina, a farm girl, goes about a normal day, doing her chores and playing with a homemade doll. Then the action starts. An automobile breaks down near the farm, and her father helps the hapless townsman repair his vehicle. Grateful for this help, the city slicker returns with a store-bought doll as a present for Christina. In the final segment, Christina tries to play with the wonderful store-bought doll but discovers that it is fragile. She ends her day going to sleep (as parents hope all children's books end) with the store-bought doll displayed in her room and her old doll tucked beneath her arm. *The Store-Bought Doll* promotes a rural notion of innocence to be preserved against onslaught from the alluring, but false and brittle pleasures of the town. Christina's life attains its value from its natural rhythms and from its integration into a meaningful whole.

Carol Thompson's *My Big Farm Book* is a book that purports to show what farm life is like in 1987. It is remarkable how little her version differs from that of Boyd Smith, published originally in 1910. The book stresses daily and annual work cycles. Family activities change with the seasons. Each member of the family has special talents and special jobs, and all go together to make the farm work. The theme is explicit in *My Big Farm Book* and implied in *The Store-Bought Doll*, where passages describing Christina's typical day show her performing a variety of functional chores. In both of these books animals appear as a matter of course. The farms are diversified. They have cows, pigs, chickens, ducks or geese, dogs, cats, and, of course, horses. Both books reinforce the key themes of the agrarian ideal. The farm is a place where men and women, young and old, have special roles to contribute.

The stories are in no sense *about* the animals, nor are the animals anthropomorphically portrayed. Animals appear as a matter of course in both books, as partners in the farm, and significantly, as parts of the farm operation for which the children bear responsibility. Neither book disguises the fact that these animals are on the farm for a pur-

---

1. I selected *The Store-Bought Doll* from my children's library with little more than a cursory search for stories that rely upon farm and animal images. Someone should do a systematic survey of recent children's books to determine what percentage of them have farm settings or depend upon animal imagery, but lacking this information, I simply don't know how typical this book is. Having spent the last decade reading books to my children, it strikes me as more maudlin than some in its moral message, but very typical in its use of imagery.

pose; indeed, the fact that everything has a purpose is essential to the agrarian ideal. Although neither book contains scenes in which animals are slaughtered, they do not conceal the fact that sometimes a farm animal's purpose is to be eaten. The books place farm animals squarely within an agrarian ideal, and it is in communicating the ideal that they communicate values for understanding animals. *My Big Farm Book* makes some pretense of showing what farming is like, but *The Store-Bought Doll* does not even do that, setting the story in another century. They establish an ideal type, a picture of how life might be lived, and how a certain framework of values emerges in living life that way.

In eschewing anthropomorphism, these two books are quite different from child literature which, like *Black Beauty*, attributes rational concepts, interests and conscious intentions to animals. In setting the portrayal of animals within the context of farm life, *My Big Farm Book* and *The Store-Bought Doll* place animals into agrarian ideal. This vision of animals is different from one which extends fundamental moral categories such as suffering or interests to animals. Relative to a children's literature that commonly casts bears, lions and rabbits in obviously human roles, these books offer highly realistic portraits of farm animals. Children may be getting a romantic view of agriculture from these books, but they are not getting overly romanticized images of animals.

It is fairly common for people in agriculture, and especially agricultural scientists, to complain about highly romanticized portraits of animals and to attribute much of the impetus for the animal rights movement to the fact that such portraits are prominent in the minds of the urban population. This assessment may well be correct, but it overlooks the fact that even when urban children get a fairly realistic portrayal of farm animals, it is a portrayal that presupposes an agrarianism that is a sharp philosophical contrast to utilitarian agriculture. The agrarian ideal presupposes a strong moral commitment to the human use of animals, but it defines the value of animals in moral terms that place more stress upon human conduct and the formation of character than upon trade-offs or exchanges made in an optimizing calculation. If the agrarian ideal (as I have described) does influence attitudes toward agriculture, it presents an alternative to the debate between utility and rights that has dominated the discussion of animal agriculture. Although a full discussion of how agrarian philosophy contrasts with individualist philosophies of utility and rights would take the present discussion well beyond its assigned scope, it is appropriate to conclude our discussion of animals in the agrarian ideal by returning to the contrast between utilitarian and agrarian agriculture with which I began.

Evaluating Animal Agriculture

A utilitarian philosophy of agriculture evaluates agricultural practice in light of its consequences. Applying the utilitarian maxim (to do the greatest good for the greatest number) to the question of which animal production system to use, one attempts to weigh the benefits and costs of various options. If, for example, one compares a food production system of diversified farms with one of concentrated and specialized production units, one will want to know which produces the greatest benefit in terms of food, when weighed against costs. The common sense of this approach conceals the fact that there are important philosophical questions to ask about how one counts the benefits and the costs. While economists are generally willing to include emotional stress suffered by humans as a cost, I know of none who have attempted to include stress to animals as part of the cost accounting in their assessment of animal agriculture. Whether animal suffering *should* be included has come to be the first philosophical question raised in the ethical assessment of animal agriculture.

Even if animal suffering is included as a cost, it is quite possible that the benefits to humans will outweigh it; yet it is clear that we would not allow human lives to be sacrificed for purposes comparable to the eating of meat. Comparable trade-offs among humans are barred when the costs for even a single individual are unacceptably high. This fact of our moral practice suggests that rights held by individuals trump the aggregate good, providing a constraint upon applications of the utilitarian maxim. If we are willing to extend our cost accounting to include animal suffering, it is not obvious why we should not extend our absolute prohibitions of some practices to include animal rights. To start with a utilitarian philosophy of agriculture is to establish a burden of proof which points us toward questions of whether animals suffer, how much they suffer, and whether they have rights. These are, of course, precisely the questions that have been taken up by Peter Singer and Tom Regan. Current animal production systems can be evaluated only when these questions of impact upon animals have been answered, either through empirical research or through philosophy (and most likely through a combination of both).

In agrarian philosophy, farming is a way of life. Its value comes from its ability to frame and reinforce a series of character-building values. Farming promotes citizenship because it forces one's attention on local community issues. Farming promotes industriousness because it is, in the agrarian world, manifestly hard work. Farming promotes family loyalty because the family and the production unit are thoroughly integrated. Stewardship, too, emerges from the family's need to attend to natural cycles. Family farms promote self-reliance because all of these virtues are knitted into a whole, a way of life that establishes and promotes a fulfillment of one's constitutional excellence. It is the whole framework that establishes value. Animals are, historically, a fundamental part of this framework, both in the role that they play and in the responsibilities that they entail. Additionally, animals provide examples of innocence, possessing an instinctual ability to fulfill the potential of their constitutional potential.

The agrarian ideal provides a way of evaluating agricultural systems that synthesizes many values, and in doing so fails to provide unequivocal evaluative criteria. While utilitarian philosophy allows us to make sense of a comparison between, say, slotted and meshed floors for hog pens, the agrarian ideal does not provide a level of specificity that will admit this as a meaningful comparison. Interestingly, the agrarian ideal thus presents an alternative not only to utilitarian philosophy of agriculture, but to the philosophy of Singer and Reagan, as well. The agrarian ideal shows how agriculture is constitutionally related to us as individuals and as a society. The ideal is communicated primarily through imagery, though it admits of philosophical articulation and defense. It is only on the level of imagery and very broad generalizations that it entails judgments about whether one production system is better than another.

Even on a broad level, however, it cannot be doubted that a concentrated and specialized monoculture is far from the agrarian ideal. Most farm families today buy their food at the grocery store and depend upon cash income in exactly the same way that city folks do. Many farm families depend upon off-farm employment for at least one member of the family. Although there is a great deal of variation from one farm (and one region) to the next, few farms unify a diverse set of family roles in the manner prescribed by the agrarian ideal. While the transition away from the agrarian ideal in agriculture has been ongoing for some time, it is only recently that the images of agriculture have started to depart from the pictures in children's books. Nowhere is this departure more striking than for animal agriculture. Confinement production systems just don't look like the pictures in *My Big Farm Book,* not to mention *The Store-Bought Doll.*

Does this divergence between the agrarian ideal and the reality of agriculture spell trouble for agriculture, or does it simply indicate the anachronism of the agrarian ideal? This question, like many others I have posed, goes well beyond my present scope. The agrarian ideal is clearly anachronistic, particularly (and not surprisingly) with respect to its formulation in Jeffersonian and Emersonian terms. On the other hand, a philosophy that evaluates agriculture solely in terms of trade-offs between cost and benefit neglects the importance of agriculture for our understanding of human conduct, character and for the constitution of democratic society. This is a tragic loss, one which American society cannot expect to suffer lightly.

11

Constitutional Values and the American Food System

Previous chapters of this book have taken up specific problems in agricultural research, teaching, and public policy. This chapter and the conclusion initiate a turn toward a more comprehensive philosophy of agriculture. There are two themes to this philosophy, as I see it. This chapter launches an expansion and revision of the agrarian values that have been associated with Thomas Jefferson and that have been exposited so brilliantly by Wendell Berry. Several of the previous chapters have made reference to Jefferson and to Berry, but bringing these essays to closure demands a more focused (and more critical) look at agrarian themes. However, agriculture (especially American agriculture) will never again be characterized by the uni-dimensional vision that we associate with agrarian themes. Hence the second task: to articulate a vision of an ideologically diverse yet sustainable agriculture, and to find the philosophical language that integrates the sciences with the educative and cultural dimensions of this vision. This latter task, begun with the discussion of a policy framework in Chapter 7, will be revisited in the conclusion of the book.

The tension between these two tasks cannot help promoting a feeling of ambivalence and a lack of closure regarding the topics considered throughout the book. On the one hand, we celebrate and expand upon agrarian themes, rethinking them in contemporary, communitarian terms. On the other hand, we withdraw from our commitment to those themes, advocating a neutral procedure for amelioration of conflicts. Though readers may find it disconcerting, I cannot regard this as a flaw. It has never been my intention to produce a comprehensive or totalizing vision of agriculture or of ethics. However, if we are to live in a diverse world we must be willing to find ways of peacefully and rationally negotiating with one another when our diversity prompts conflict. In this vein, the last chapter will take up the conflicts arising from incompatible ethical perspectives and will do so without endorsing any perspective. But a purely procedural agricultural ethic will not do precisely because it is quite unlikely to appreciate agriculture's unique contributions to our valuespheres. Better decision making about agriculture requires both a commitment to a procedural politics that treats agrarian ideas as just one of several competing visions for agriculture, *and* a willingness to reexamine and rehabilitate the traditional values of the agrarian past. It is in that sense that the present chapter takes up the theme of constitutional values.

Constitutional and Other Values

As noted in the first pages of the introduction, the word "values" is both indispensable to ethical philosophy and a persistent source of confusion. This point came home to me when I was asked to speak at a symposium on "calculating the true costs of food" sponsored by the Wallace Institute for Alternative Agriculture. The general idea behind the symposium was clear enough: though we think that food is cheap, if we better understood the human and environmental costs of producing food the way we do it at the close of twentieth century, we might revise that judgment. Although I wholeheartedly endorsed that theme at the outset, I found myself becoming increasingly uncomfortable with it. The "true costs" that were being calculated involved resource depletion, pollution and human psychological stress, all important and worth considering, but also types of cost that were fully compatible with the utilitarian way of thinking about agriculture that has been a frequent topic of critique in previous chapters of this book. There are better ways to think about values.

First, however, it is important to give these utilitarian notions of value their due. Agricultural economists typically calculate costs in terms of exchange value. The cost of a good can be evaluated by identifying the other goods that must be forgone in order to acquire or retain it. It is easy to identify cost with price, a value that is more completely determined by people's willingness to exchange one good for another. Critical economists such as Harold Breimyer (1965), Al Schmid (1987) and Daniel Bromley (1989) have shown why market prices are sometimes a poor indicator of true costs. When hidden costs, transaction costs and opportunity costs have been factored in, value can be more faithfully represented as a willingness to exchange one bundle of goods for another. With respect to food, this means that we frame the evaluation of our farm and food policies as a question of whether the relatively low percentage of consumer income spent on food in the United States is worth the cost in government farm programs, degradation of soil and water resources and perhaps aesthetic or health impacts that are the result of chemically intensive production and processing. The range of exchange values can even be expanded to include existence values of rural communities.

Determining these values is not easy. The new generation of ecological and resource economists is to be commended for innovating theoretical and empirical methods to achieve a more faithful and complete calculation of these costs. In other writings I have applauded their efforts and join them in resolving the conceptual, epistemological and methodological difficulties that must still be overcome (see Thompson, 1995b, Chapter 5); the point of this chapter is somewhat different. For, despite the impressive advances in calculating the true costs of food, exchange values are *not* constitutional values. There is still more to be said about the future of U.S. agriculture, the food system and, in turn, American society.

For most Americans the word "constitution" refers to the document housed on 7th Street and Pennsylvania Avenue in Washington, D.C. More broadly, constitutional choice is thought to involve procedural questions about the division of powers or about the imposition or removal of constraints upon the various branches of government. In some quarters, constitutional values refer to the original intentions of founding fathers who gathered to draft the United States Constitution. Since food and agriculture are little discussed in the U.S. Constitution or the Bill of Rights, one might wonder how constitutional issues could figure in a review of the American food system.

Outside of the United States, however, observers of society from Rousseau to Anthony Giddens (1984) have used the word "constitution" to indicate cultural values that define a nation, that determine a people's practice and that shape social conduct in the most fundamental and enduring ways. Rousseau's references to the constitution of society are scattered through his writings. Reference especially *The Social Contract* (1762, trans. 1984) at 135, where Rousseau has been translated:

> The constitution of a man is the work of nature; that of the state is the work of artifice. It is not within the capacity of men to prolong their own lives, but it is within the capacity of men to prolong the life of the state as long as possible by giving it the best constitution it can have.

Unlike exchange values, and even unlike the procedures of constitutional choice, constitutional values are the values that define American culture and society. The U.S. Constitution is, of course, deeply relevant to the constitution of American society in this sense; but so is agriculture.

Agriculture and Constitutional Debates in American History

Agriculture figured in debates over the U.S. Constitution during a dispute between more democratically inclined founding fathers, such as Thomas Jefferson, and the Federalists, such as Alexander Hamilton (Thompson, 1988). Hamilton advanced the view that the new republic needed to encourage the growth of a class of enlightened and public-spirited leaders who would be able to dissociate personal and public interest. Like Edmund Burke in England, Hamilton had no hope that persons with little education or experience in statecraft would ever be persuaded to set aside the pursuit of narrow self-interest, or would make the sacrifices necessary for achieving a stable and just society. Madison advocated a federal system with checks and balances not only among the branches of the federal government, but within each branch and between federal and state authority. Madison hoped that sheer complexity could dampen the fires of public passion.

The Federalists were expressing an idea that had been commonplace in discussions of political thought since Plato, and which was to be a central theme of Alexis de Tocqueville's study of American democracy a few decades later. It is the idea that citizens of a democracy are likely to insist upon government serving short-term private interests even when it sacrifices the society's long-term survival. Put even more simply, it is the fear that democracies will vote themselves benefits but will defeat the taxes necessary to defray the costs. The Federalists' fears are a cogent legacy for any group assembled to discuss the true costs of food in America today.

As noted in Chapters 6 and 10, Thomas Jefferson's writings on agriculture must be read with reference to this constitutional debate. Jefferson was an advocate of democracy, and he countered Federalist fears by turning an argument that had been used to defend aristocracy upon its head. One reason for investing power in a landed aristocracy had been that since the income and status of aristocrats depended upon the security and stability of their lands, they would be far more judicious when considering affairs of government than would artisans or tradesmen, whose assets were moveable and could be transported from state to state (Berry, 1987, p.144). In passages from a celebrated letter to the Federalist John Jay (see Chapter 5), Jefferson noted that in America, it was the common citizen, the free-holding small farmer, who occupied that land. If European aristocrats could identify personal and public interest in virtue of their landholdings, then so could American farmers (Jefferson, 1984, p.818).

In fact, eighteenth-century political theorists were preoccupied with agriculture in a way that is difficult to imagine today. Historian Garry

Wills traces many of Jefferson's remarks on agriculture to one of his French contemporaries, the Marquis de Chastellux, whose treatise on public happiness was published in 1772. Chastellux reduced all positive contributors to human happiness to two. One is simply the number of people (the more the merrier), but the other is agriculture. As Wills reports it, Chastellux believed that "Farming is the ideal state for man. It is the golden mean placed between wandering like nomads and being packed into corrupt cities," (Wills, 1978, 160). Combining Chastellux's two principles, one finds that the nation is the happiest that manages to support the largest number of farmers relative to the whole population.

Of course, both Jefferson and Chastellux were participants in a conversation rich with tradition in the Enlightenment period. Agriculture and farming practice figures prominently Locke's 1690 theory of property, which defends the practice of enclosing common lands for farming. Enclosure had been opposed in England by Gerrard Winstanley in a 1649 treatise that also extolled the role of farming in establishing the basis of civil society. At about the same time (1656) James Harrington published *Oceana,* which argued that "He who controls the land will control the supply of food, he who controls the supply of food will control the army, and he who controls the army will control the government" (Montmarquet 1989, 40). From this Harrington deduced that a king must control the land, or that a commonwealth would succeed only in conjunction with a wide dispersal of landholding, since only then would commoners feel that military service was consonant with their private interests.

Across the English Channel François Quesnay was publishing the economic theories that became known as "physiocracy." The key physiocrat doctrine is that through agriculture and only through agriculture is there any genuine increase in wealth. Quesnay's doctrine is now viewed as economic claptrap, but it's political implications are what is of interest here. Quesnay was to use his doctrine to argue for the reform of a system of taxation that was stifling the growth of agricultural productivity in France. French taxation was not far removed from simple patronage, where each farmer kept only that portion of the crop they needed for sustenance, owing the rest to the patron. The system both reduced the incentive for increased yields and married itself to a method of securing revenue that was bound to grow at a much slower rate than the economy in general. In this context, Quesnay's arguments were aimed at releasing French agriculture from the fetters of a regressive system of taxation.

Despite their limitations, Harrington and Quesnay are excellent examples of political thinkers who see the philosophy of agriculture in constitutional terms. The centralized power of the French throne affords more control at the outset; there is little question that the French monarchy had a firmer grip on affairs than the English crown during the late 1600s. This is just as Harrington expected, but in the long run looser control of agriculture means both more economic growth and greater loyalty. Quesnay is vindicated. The English experience their rebellions early, but unlike France, the Crown survives (Goldstone, 1991). By the late eighteenth century, both Jefferson and Chastellux were in a position to observe the long-run constitutional implications of farm policy at close range.

Thus, Jefferson's comments on farming are no anomaly in the political theory of his time, and like Harrington or Quesnay, Jefferson's views are relatively undeveloped. Much of what we in the twentieth century need explained about farming and constitutional value was simply common knowledge for any educated person of Jefferson's time. Yet there is a dimension to Jefferson's argument that is lacking in others'. Borrowing from Harrington, Jefferson sees yeomanry as crucial to democracy, but his view is dynamic and developmental, and it extends well beyond the need to raise armies. In this he anticipates Kant, who includes several long paragraphs on the transition from herding to agriculture in his 1786 essay, "Speculative Beginning of Human History." Writing at about the same time as Jefferson's letter to Jay, Kant's words sound remarkably Jeffersonian to the American ear:

> Since the herdsman leaves behind nothing of value that he cannot find in other places, it is a simple matter for him to move his herd to distant places, once he has done his damage, and in so doing he avoids having to make good on the damage. ... When subsistence depends on the earth's cultivation and planting, ... permanent housing is required and its defense against all intrusions requires a number of men who will support one another. ... Culture and the beginning of art, of entertainment, as well as of industriousness (Gen. 4:21-22) must have sprung from this; but above all some form of civil constitution and of public justice, (Kant, 1983, p 56).

Jefferson's reply to Hamilton's concern was a political and philosophical master stroke, but that it anticipated Kant, indisputably the greatest philosopher of his generation, is greater testimony to Jefferson's genius.

The farmer was valuable *as a citizen* because he would be well-placed to see the convergence of personal interest and the public good. The idea of the public good at work here is the stability and fiscal responsibility of the society. The farmer's loyalty to the state is a constitutional value for both individuals and society in that it is taken as a defining characteristic of citizenship. It is useful to think of citizenship as a role for which one might prepare as an actor studies a script. It was clear to all eighteenth-century political thinkers that some people would be better equipped for the part than others. Each member of society's ability to play the role of citizen hangs upon having sympathy with the part, on being able to motivate oneself for faithful conduct of the citizen's responsibilities. Thus performing as a citizen requires a certain kind of moral character. Jefferson argued that in America freeholding yeomen farmers would have the required character because their farming activity would make it obvious that the state's ability to protect their lands served personal interests. Because they could not threaten to remove themselves or their farms they were required to accept the need for resolving political differences in ways that did not threaten state stability.

There are, in effect, two points here. One is a derivation from Harrington's coldly realistic observations. A state will be strong to the extent that the interests of its citizens are married to the ideal of stable government. But there is nothing in this that would create development in the character of citizens beyond self-interest, much less a historical development tending toward democracy. The second point sees interest as the fuel that propels individuals and societies alike through a developmental process that leads to a fuller realization of both personal and collective moral potential. When authors like Rousseau or Kant spoke of an individual's constitution, they meant more than the person's general state of physical health. Character traits are thought to be fundamental components of one's constitution. The process of education or enlightenment turns the mean into the charitable, the jealous into the kind. For individuals, constitutional values indicate character traits that manifest themselves in longitudinal patterns of conduct. The constitution of a society could be understood as an aggregate of individual constitutional values, and it, too, would become evident through its history.

The problem of constitutional choice is that of finding a way to bind individual patterns of conduct to one another in a functional manner. Jefferson's argument was intended to suggest that farming would induce long-running patterns of conduct that would be broadly characteristic of American society as a whole. A wise government would not only rely upon the character of its citizens as a source of social unity but would be attentive to government's impact upon the patterns of conduct that form character in the individual. When he was young, Jefferson thought that the American republic would mature in a way that would allow merchants and manufacturers to share in this vision of social unity, but as he grew older, he became even more strongly committed to the belief that life on the land was essential to citizenship (Wills, 1997).

This developmental theme in Jefferson's thought took root in the romantic agrarianism that runs throughout the writings of Ralph Waldo Emerson in the nineteenth century. As noted at some length in Chapter 5 and again in Chapter 10, the romantic development of agrarian thought introduces at least two new themes. First is the notion that nature is a formative element in the American national character, second is the related idea that hard physical labor is a prerequisite for achieving the flowering of virtues necessary for self-realization.

Emerson's work on nature was explicitly linked to his concept of moral character. For Emerson, realization of one's potential for appreciation and expression of value was the overriding purpose of human life. The fullest realization of human possibility consisted in poetic creation, for Emerson, and the appreciation of nature was a vital component of the creative process. In the essay *Nature* (1965), Emerson writes that the human spirit has been fitted for the sights and experiences of the natural world. The artificial world of the city forces people to work in an environment that is not conducive to the realization of natural powers. Like Plato's cave, the city is a world of broken time and shrouded light, a place in which the human impulse to bestow meaning upon the common place is apt to fix upon transient or ill-formed objects of attention. While apparently simpler than city life, the world of nature is, for Emerson, a place of infinite detail and richness, well-suited to the human impulse for inquiry, representation, and explanation (Corrington 1990).

While Emerson shared this vision of nature with Henry David Thoreau (see Thoreau, 1965), he differed from Thoreau in seeing nature as a place where human beings were busily at work transforming the natural landscape. This meant that farming and work were far more crucial themes for Emerson than for Thoreau. For Emerson, nature was not merely an object of contemplation, it was the arena in which the human drama was enacted, and the material that must be transformed if one is to progress in the realization of self that is mankind's purpose on earth.

The problem is that nature must be transformed in ways that are authentic expressions of human spirit, that ring true to the innermost core of being. This problem is most acute for the artist or poet who aspires to create but who lacks the vision and instincts to realize a true work of art. Aiming to please the critic, the false poet creates not from a strongly felt grasp of place, but by imitating forms that have won the approval of others. In comparison to the false poet, the farmer is much closer to a realization of the human ideal, despite a relative lack of formal education. The farmer cares little for the opinions of others. The farmer must create in one lifetime a work of art whose only judge is nature. If nature judges the work well, lifetime farmers thrive; if not, they suffer. Given that they work in and with nature, farmers develop traits of observation and invention that are closely tailored to humanity's evolved capacities of sensation and ratiocination. For this reason also, the farmer's life activity is a far more authentic production of the creative spirit than is the false poet's imitation of art (Corrington, 1990). The farmer, thus, is a signal indicating the direction toward spiritual fulfillment for even the true poet, who will be better-served by observing farm life than by studying great books.

Emerson's works develop the link between farming and aesthetic character just as Jefferson's stress the link between farming and citizenship, or political character. The triangle of farming, citizenship, and character is strengthened by Emerson's work because he provides a plausible way of understanding how farming contributes to the achievement of moral virtue. The good person follows the path of the true poet, aiming toward authentic realization of self. Most will not reach the poet's vantage point on nature, and no one can remain there indefinitely. The Emersonian vision was carried forward by several generations of American intellectuals, reaching its flower in John Dewey's belief that American democracy realizes its creed when the path of the true poet is open to all, where all can realize an inner potential for expression and creation. (Dewey would have been more likely to celebrate the creative aspects of science than of poetry. See Dewey, 1939.) According to this romantic vision, a democracy of farmers provides a land rich in signs that point the way toward self-realization, and forms a community of individuals who actualize human potential far more fully than any society in human history, including ancient Greece.

Constitutional Values and Exchange Values

It is obvious that a transition has occurred that makes these Jeffersonian and Emersonian themes seem like quaint relics with little relevance to the political life of the twenty-first century. For one thing, few of us experience the character-building experience tied to farming because we do not farm. For another, those who do farm increasingly tend to see their operations as a business and resent the suggestion that they should be held up as moral exemplars. But how should we evaluate this change? Early on in this chapter, I defined constitutional values in contrast to exchange values. Exchange values are determined by trading bundles of goods; constitutional values are determined by character traits that underlie long-run patterns of conduct (though they need not express themselves as conduct in any particular case). There are occasions on which constitutional values cannot be realized, and if we take the Jeffersonian and Emersonian statements on agrarian life at face value, our own time may well be one of them. It is possible to interpret this as evidence that constitutional values are subject to trade-offs and to infer that a loss or deterioration of constitutional values can be treated as a cost, as a value given up in exchange for something else. In my view, such an interpretation makes a highly metaphorical use of the term "cost." Individuals incur costs from their conduct and decisions to the extent that goods and opportunities having value to them are denied. A change in one's constitution is more fundamental in that the entire framework of conduct and choice is potentially transformed in ways that suggest a very different person is now acting or choosing.[1] A properly specified notion of cost shows how, in any act or choice, valuable goods and opportunities are being relinquished in pursuit of other valuable goods and opportunities. Describing change in constitutional values as a cost implies that there are some further background criteria for valuation which permit comparison between being one kind of person (or society) and another. As a metaphor, talk of such comparisons is very useful, but the suggestion that the trade-off between being one kind of person or another can be represented in one's willingness to exchange material goods overlooks the fact that it is as the person one is today that such choices must be evaluated, and one cannot adequately know how the person one is to become would evaluate them. This is not to deny the possibility of interpersonal comparisons of utility, but rather to say that comparing the prospect of being one kind of person with the prospect of being someone rather different is necessarily a philosophical or literary task and not one that lends itself to exact measurement.

The philosophical problem I am alluding to is the Faustian bargain. A naive view of a devil's bargain assumes that the foolish youth trades

1. See Parfit, 1984, for a discussion of how such transformations in personal identity are related to problems of choice.

a moment of bliss for an eternity of torment. The deeper philosophical lesson is that anyone who makes a bargain to buy now and pay much, much later neglects the incompatible constitutional values that determine both the worth of what is given and what is gained. The devil knows that the individual develops and deepens over time, in effect becoming a different person, a person who applies a wholly different framework of valuation to those prior choices. In the classic tales of the Faustian tradition, the callow youth is presented with an irresistible choice, for the devil knows that a person's growth will transform character over the long run.

Applying the Faustian bargain to America's choice of food systems requires a double metaphor. Historical change in constitutional values is only metaphorically a "bargain" because, as with Faust, the party that makes the contract is not the party that must live up to its terms. Furthermore, the party making the choice is only metaphorically a person, being instead the aggregated character traits of a society of individuals. We are attempting to compare the possibility of being one kind of society against the possibility of being a different kind of society. If we bear these two metaphorical translations firmly in mind, we will be able to resist the temptation of thinking that we can confidently quantify (or worse, discount) trade-offs at the constitutional level. Bearing the metaphorical nature of the topic in mind, we can talk about our current dilemma as a transition in constitutional values. And we can also see that some of the choices we made in the 1960s and 1970s were also a devil's bargain.

As such, while it is true that the constitutional values indicated by the agrarian triangle must sometimes be sacrificed, it would be mistaken to describe them as values that are set by exchanges on goods to be ranked along with television sets, clean water, and tasty food. The sacrifice of agrarian constitutional values may be necessary, but when it occurs it is a tragedy, not a trade-off. The loss of constitutional values cannot be calculated among the true costs of food, for if they are truly gone, then we are a vastly different sort of people than our ancestors, and the map provided by Jefferson and Emerson points toward a very different America than we are heading for.

Heirs to a Faustian Bargain

The society that was America faced a choice sometime after World War II. Lured by the diversions of an urban culture, Americans became fascinated with technologies and institutions that transformed rural life, while producing an abundance of food and fiber for consumption by an increasingly urban society. In effect, the choice was between two paths for alleviating the menial and unfulfilling aspects of food production. One path was that of Jefferson and Emerson, to produce food by first producing a democracy of citizens whose characters were infused with the agrarian virtues of community spirit, stewardship and hard work. The other path was to develop technological substitutes for these virtues. While the technological option may have appeared to be an attractive choice to the society that made it in the 1950s, we find ourselves as a very different sort of people in the 1990s. Now we want a food system that contributes to our constitutional democracy, but we have become so addicted to our material abundance that we must trick ourselves into changing our ways by concocting stories about the true costs of food. We are no longer a nation of farmer/citizens, bound to a democratic community through character traits of work and the appreciation of nature. We are instead a nation of shoppers, evaluating our conduct as a series of consumption decisions. We are a nation of consumers who compare policies the way we compare breakfast cereal. We have lost the ability to see public policy as laying out a framework of values that will affect our future patterns of conduct, and which will determine the character of generations of Americans yet to come.

Current policy also makes it hard for farmers to think of themselves as stewards, much less the rest of us who purchase food in packages that must be carted from our homes in garbage trucks to be deposited, along with an alarming amount of food waste, in land fills and garbage heaps. Calculating the true cost of our food is beginning to show consumers how nature is judging our practice, but so far we have responded in ways that only frustrate farmers' ability to see themselves as stewards. According to a *Farm Futures* poll, more than half of the respondents felt the ethical standards of producers had slipped over the past decade. On some matters, up to one-third admit to unethical practices in their own operation. The poll suggests that farmers feel besieged by regulation in ways that preclude them from acting in conformity with the law (Knorr, 1991). I suspect that most farmers would respond to this problem by eliminating regulations, rather than by reforming policy to more effectively influence practice toward the norm of stewardship. Nevertheless, we must see the challenge as one of finding a policy that forms character, as well as internalizing environmental costs. That is, ultimately, what sustainable agriculture must mean.

Conclusion

Markets, Moral Economy, and the Ethics of Sustainable Agriculture

The previous chapters review a series of issues in agricultural research, teaching and the broader public understanding of an institutional structure of farming and food systems—what I regard as an updated version of the triad long thought to constitute the land-grant mission. This concluding chapter provides a sketch for a more comprehensive philosophy of agriculture which I call (with no great originality) *sustainable agriculture.* As with the general organization of the book, it touches on research, teaching and public policy. In fact, returning to some of the themes discussed in the first three chapters, it is research that turns out to be crucial. I will argue that we do not have an adequate conception of sustainability as it might apply to agriculture, and that philosophy (including moral philosophy) will play a crucial role in conceptualizing sustainability in a researchable manner. In making this argument, I will say how my conception of sustainability relates to several identifiable moral positions in contemporary philosophy, but I will not try to position my argument as a form of philosophical communitarianism, utilitarianism, Kantianism or any of the other "isms." Attempting to bring the best reasoning one can muster to bear on the question of sustainability is more than enough of a task for a concluding chapter, and those who would expect a philosophy of agriculture to express one version of a particular moral vision will have to look elsewhere.

Reviewing the literature on sustainable agriculture in 1984, Gordon Douglass found three patterns of conceptualization. One group of authors utilized a *resource sufficiency* conception of sustainability. On this view, a practice is said to be sustainable over a given period of time only if the resources needed to carry on the practice are on hand or foreseeably available. A second group stressed *ecological sustainability.* On this view, a sustainable practice is one that does not violate or disrupt natural biological processes, especially when biological processes are essential to renewal of the organic materials necessary for life. The third group of authors stresses *social sustainability.* Douglass characterizes these authors as concerned with justice and equal opportunity. Advocates of social sustainability seem to be saying that social goals should not be sacrificed at the altar of resource sufficiency or ecological sustainability (Douglass, 1984, 14-18).

My ultimate goal in this chapter is to provide a more convincing conception of social sustainability, but first it will be useful to analyze why these three patterns of conceptualization are sometimes interpreted as competing paradigms. Here I use the term "paradigm" in the sense popularized by Thomas Kuhn (1970) and applied to environmental policy analysis in a recent paper by Bryan Norton (1995). Norton defines paradigm as "a constellation of concepts, values, and assumptions, as well as accepted practices, that give unity to a scientific discipline." (p.113). To that I would add that a paradigm includes an implicit specification of unsolved problems and key research needs. My main thesis is that while *social sustainability* is essential to an ethic of sustainable agriculture, it has not been conceptualized in a manner that implies a clear research agenda. For this reason if no other, the dimensions of social sustainability have been neglected over a decade of research on sustainable agriculture and sustainable ecosystem management.

Paradigms of Sustainability: Resource Sufficiency and Ecological Sustainability

Resource sufficiency defines sustainability as an accounting problem framed by two key value questions. What practices do you want to undertake, and how long do you want to undertake them? Given answers to these questions, one can perform research to identify both existing and optimal rates at which resources are needed to support the specified practice, and from this one can calculate the total amount of resource needed. Alternatively, one can measure both the rate of resource consumption and the existing supply and calculate from these measurements how long a given set of practices is sustainable. Resource sufficiency has clearly been especially influential in conceptualizing sustainable development in a manner consistent with the Brundtland definition of 1987, "development that meets the needs of the present generation without compromising the ability of future generations to meet their own needs." (World Commission on Environment and Development, 1987, p 43)

Although it may seem absurd to specify an infinite time horizon for resource sufficiency, it is possible to conceptualize practices as sustainable into the indefinite future by determining resource sufficiency over a definite but rolling time horizon. To do so, one presumes that ever-increasing efficiency in the utilization of non-renewable resources results in continuous decrease both in the present consumption of the non-renewable resource, and in the amount needed to maintain resource sufficiency over each new iteration of the rolling time horizon. This approach to sustainability thus specifies an agenda of research in resource utilization that is reasonably clear and of obvious value to resource managers. Henk van den Belt ties a Brundtland-type conception of sustainability to the Dutch debate over measuring the environmental utilization space (EUS). According to van den Belt, the concept of an EUS deploys several elements associated with ecological sustainability (discussed below) but takes a resource sufficiency approach to non-renewable resources (van den Belt, 1995).

Van den Belt's analysis of the EUS debate indicates the first of two problems that plague resource sufficiency conceptions of sustainability. Resource sufficiency is only meaningful once the key value judgments have been made. What do we want to do, and at what rate do we want to do it? When resource sufficiency conceptions of sustainability are framed for society as a whole, these two questions encompass more than two millennia of disagreement and debate over the purposes of human life and over the distribution of access to life's most attractive opportunities. As such, the allegedly objective science required for doing the accounting called for by a resource sufficiency notion is in fact thoroughly, and perhaps deceptively value-laden. The technical clarity of resource sufficiency conceals the fact that accountancy begins only when the fundamental decisions have already been made (van den Belt, 1995).

What is more, resource sufficiency places an implicit emphasis on non-renewable resources. These, after all, form the greatest challenge to extending its time dependent dimensions into the indefinite future. This emphasis may reflect the importance of *development* rather than agriculture in the Brundtland definition. Resource sufficiency provides no reason to treat resources such as water and soil as different from resources such as fossil fuel, since the same efficiency-increasing logic may be used to account for either over the period for which resource sufficiency is calculated. Yet much of the impetus for sustainable agriculture, as distinct from sustainable development, has come from the belief that it should be possible to farm in ways that do not deplete soil and water resources at all.

Ecological sustainability presupposes that human activity is nested within functional biological systems. Ecological sustainability thus frames conceptual and research questions in terms of a need to understand first the processes that renew resources, and second the impact of human activity on these biological systems. Executing this research program has proved to be more difficult than first expected, for there is far less stability in natural ecosystems than originally imagined. It may in fact be the case that human management of nitrogen cycles, soil formation, and watersheds in traditional agriculture produces more stable ecosystems than typically exist in nature. Whatever the case, ecological sustainability differs from resource sufficiency at least in that it provokes us to examine such processes and to formulate models of the systematic interaction of these processes through the median of living organisms and their life cycles. The formulation, testing, and revisions of such models becomes the leading idea behind the ecological sustainability paradigm.

Norton (1995) believes that resource sufficiency and ecological sustainability (my terminology, not his) constitute incompatible paradigms for the policy sciences. The incompatibility arises from the fact that advocates of each conception have incompatible views on the inter-substitutivity of resources. According to Norton, advocates of resource sufficiency take the neoclassical economic assumption of substitutivity quite literally. On this view, for any two goods ***A*** and ***B***, there is some amount of ***A*** such that economic agents are indifferent between a bundle of goods containing ***A***, and an otherwise equal bundle of goods containing ***B***. To say that ***A*** substitutes for ***B*** implies that when one has enough ***A***, it compensates for the loss of ***B***. Economists such as Julian Simon and Robert Solow have argued that learning (human capital formation) substitutes for non-renewable resources such as fossil fuels as the relative supply of these two goods shifts from a relative scarcity of human capital to a relative scarcity of fossil fuel. Economic jargon aside, the idea is that our efficiency in the consumption of non-renewable resources increases over time because we "substitute" knowledge for non-renewable resources. The increasing scarcity of goods such as fossil fuels is a key part of the argument, since such scarcity makes learning cheap in comparison to more consumption of fossil fuel. Market forces, thus, serve as the mechanism for substituting one good for another.

Norton shows that advocates of ecological sustainability reject the hypothesis of intersubstitutivity for resources. Nowhere is the qualitative difference between resources more important than with respect to renewable resources. Fisheries provide an impressive object lesson. As long as a breeding stock of fish is maintained, fishers may harvest a continuous flow of fish from the fishery. However, fish do not begin to become scarce (signaling a need to change human behavior through market forces) until breeding stocks have been fished beyond the point of recovery. Hence even if we accept the claim that market forces provide incentives to conserve steady-stock resources such as fossil fuels, they may fail with respect to stock and flow renewable resource pools. Advocates of ecological sustainability call for an approach to public policy and resource management that recognizes the different ways that resources can be depleted or replenished, and argue that the use of resource sufficiency tools to define sustainability for renewable resource systems may be inappropriate. Norton argues that the different assumptions of the resource sufficiency and the ecological sustainability approaches result in inconsistent policy approaches. He calls this a paradigmatic difference because, in his view, it can be traced to essential elements in the fundamental assumptions of neoclassical economic theory (Norton, 1995).

Although resource sufficiency and ecological sustainability are in some tension with respect to their key value judgments, it is not clear that they are truly competing paradigms. The problem of inadequate market signals noted above is entirely consistent with neoclassical economic theory. An advocate of resource sufficiency might well recognize the difference between types of resource for example, but would tend to reflect this difference by comparing the cost of using a renewable resource pool so as to maintain stock and flow with the cost of depleting it as one might for a truly non-renewable resource. Both neoclassical and ecological economists might attempt to resolve this problem with institutions that provide better mechanisms for informing fishers of the true cost of their practice (or, in economic jargon, for internalizing the costs of depletion). For their part, advocates of ecological sustainability must accept something very much like the idea that non-renewable resources can continue to be depleted at ever-slower rates, at least while one searches for permanent renewable substitutes. Clearly there may be disagreements about how much depletion should be allowed, but Norton has not made it clear that these disagreements are paradigmatic, rather than more straightforward value disputes. Value disputes will generate incompatible policy prescriptions, but it is simply not clear that they make it impossible for someone working within a paradigm of resource sufficiency to "see" in the same way that someone working from ecological sustainability does.

Thus while each of the advocates of these two conceptions of sustainability would be willing to subsume the other, each has some use for the knowledge produced by the other, and each leaves conceptual space for the other. An advocate of resource sufficiency will find knowledge of the rate at which resources are renewed useful, even essential, to the accounting problem, even if they see no reason to grant special status to so-called renewable resources. The advocate of ecological sustainability will recognize that some consumption of non-renewable resources will be required by virtually any scheme of agriculture currently being contemplated, and the potential for substituting non-renewable for renewable resources (as with chemical fertilizers) will exist so long as it is economically feasible. These two paradigms are scientifically compatible, even if they are linked to mutually incompatible social value judgments. Researchers may happily apply themselves to both research paradigms, blissfully ignorant of whether policy makers share their fundamental value assumptions.

Paradigms of Sustainability: Social Sustainability

As Douglass noted in 1984, a large contingent of authors writing on sustainability appear to be talking about something rather different from either resource sufficiency or ecological sustainability. Kenneth Dahlberg, for example, claims that the agronomic and agroecology work on sustainable agriculture has "a serious weakness ... it does not recognize that in the longer term it can be successful only to the degree that other portions of the food system *and* the larger society also become more sustainable and regenerative." (Dahlberg, 1993, pp. 81-82.) Dahlberg believes that a pattern of industrialization in agriculture threatens the food system. Ironically, Dahlberg's first point is endorsed by strong advocates of industrialization. A respondent to a 1990 survey of agribusiness executives' opinions on sustainable agriculture characterized it as "that level of productivity that allows the agricultural enterprise to be economically competitive ... [a sustainable practice] offers in [its] own way a savings consistent with profitability under the free-enterprise system. That means sustainable agriculture to BASF." (Richgels et al. 1990, p. 31) Clearly there is something other than resource sufficiency *or* ecological sustainability at work in these quotations.

In contrast to resource sufficiency conceptions, those who have advocated social sustainability appear to wear ideological agendas on their sleeves. One version of this ideology has stressed profitability. Farmers, ranchers, and agricultural suppliers must generate income sufficient to repay loans and purchase inputs for the next production cycle, hence production systems that fail to regenerate the capital needed for each cycle of production are not sustainable. Whatever flaws this conception has, it at least roots the sustainability of a farming practice within a process of regeneration. Since the demand for food is inelastic, there always will be capital available to support investments in agricultural production. The cost of this capital will fluctuate dramatically, of course, but in theory at least, these costs should be passed to consumers in the form of food prices. If the supply of capital is assured, the problem of sustainability is simply one of ensuring that farmers are profitable, that is, that they can indeed recover all of their production costs, including the cost of capital, from the sale of the commodities they produce. Productivity is crucial, since it simply *is* the ratio between the value of inputs and the value of outputs. On this view, profitability becomes a synonym for social sustainability.

A leftist version of social sustainability has stressed two points, summarized by Douglass as follows:

> The first is about justice—or fairness—in the relationships which develop among community members, and the second is about participation in the marking of social decisions. In the hands of alternative agriculturists, justice refers primarily to the norm of equalized opportunity for all members of a community. ... This means that social and economic structures of community life must not be allowed to create vast differences in the access of individuals to acceptable standards of nutrition, health, housing, and education, nor to bar them from full participation in the social and political systems of the community. These interpersonal differences of opportunity must be limited not only among today's citizens but also between generations, lest the present members of the community become profligate and destroy opportunity for their children. (Douglass, 1984, p. 18.)

Douglass' view has been substantially updated in the essays collected in Patricia Allen's *Food for the Future.* Allen argues that society and nature are "co-produced" and that distinctions between society and nature are either arbitrary or politically motivated. This argument permits her to interpret both resource sufficiency and ecological sustainability as conceptual tools for repressing or simply ignoring the social dimensions discussed by Douglass ten years earlier. (Allen, 1993, pp. 8-11.)

Unfortunately, both approaches to social sustainability are dead on arrival with respect to generating a research program. The rightist version of social sustainability has failed to generate a meaningful research agenda because its advocates have faith that markets will continuously regenerate profit incentives to ensure the production of food. Once markets are present, profitability becomes a necessary condition for sustainability and there is nothing more to learn. If the markets are not present, the only explanation is that government interference has blocked their emergence, hence, again there is nothing new to learn. While this overstates the situation a little it is nonetheless true that the kinds of research on market structure, finance and productivity that would contribute to a profitability-based conception of social sustainability are nothing new. Research of this sort has been conducted by farm management experts and agricultural economists for many decades.

The leftist version fails to generate a research agenda for reasons that are more subtle and that I have examined at some length in *The Spirit of the Soil: Agriculture and Environmental Ethics* (1995b). I will summarize the argument here. One can (and our forebears have) conceptualized fairness, equity, justice and the other central normative concepts for evaluating a given civilization without reference to its sustainability. The most plausible way to develop a normative position that reflects our new knowledge of environmental threats to the future is to say that a civilization should be sustainable *in addition to* being fair, equitable and just. This approach poses a philosophical puzzle in that we must ask when it might be appropriate to accept a decline in fairness, equity or justice in order to have more sustainable society. Indeed, this is a familiar question for some who have advocated either resource sufficiency or ecological conceptions of sustainability.

The left-oriented advocates of sustainability challenge the validity of this question by claiming that unfair, inequitable and unjust practices make a society unsustainable. Lori Ann Thrupp asserts this, for example, in her contribution to the Allen volume when she writes that "the most important causes [of land degradation] include the inequitable control of resources, short-term economic interests and resource exploitation, and skewed policy incentives (i.e., state influences) embedded in the prevailing patterns of uneven development." (Thrupp, 1993, pp. 53-54) Yet this is an impossible proposition to prove with social science research. The "skewed" incentives are only one component of the social environment that structures the opportunities of economic agents and it is the totality of this structure that causes land degradation. Proving the even broader claim that inequality is itself unsustainable would require that one eliminate every repressive and unjust alternative to the status quo from the universe of potentially sustainable societies. One can readily imagine political regimes with food production systems that are sustained through systematic exploitation of forced labor and even through ruthless execution of dissenters. There are arguably historical examples of such systems to be found as well. If one cannot *prove* that repression is unsustainable, the alternative is simply to *define* sustainable agriculture as agriculture that meets the criteria of fairness, justice, participation, etc. But here one seems to be using the word "sustainable" as a substitute for "morally acceptable." Why not simply say morally acceptable agriculture?

Allen and her contributing author are on firm ground when they point out that programs for sustainable agriculture have overlooked the problems of hunger, gender, minority rights, and unfair labor practices. (Allen and Sachs, 1992, 144-150.) When advocates of sustainable agriculture propose solutions to *environmental* problems that fail to recognize these *social* problems, they leave themselves open to valid moral criticism, but the advocates of social sustainability have not offered convincing accounts of the mechanisms that link these moral problems to the non-sustainability of a social system. This failure is most evident in philosopher Tom Regan's contribution to Allen's book. Regan runs through the standard list of philosophical approaches to ethics, examining why and whether they entail vegetarianism. Regan concludes that despite a diversity of conceptual approaches, all arrive at a rejection of so-called factory farming, thus ethical vegetarianism is a component of sustainable agriculture. (Regan, 1993, 103-121.) Now there is clearly a basis for being concerned about shifts in animal production that replace extensive animal grazing (where nitrogen cycles and energy use is low) with intensive or confinement systems that utilize more fossil fuel and turn nitrogen into a pollutant, but these concerns emerge out of an ecological conception of sustainability. Regan's discussion of vegetarianism adds nothing to ecological sustainability. If there are morally compelling reasons to become vegetarians, these operate independently of how we understand an agriculture to be sustainable or not. Incorporating these norms into one's definition of sustainability burdens the word with rhetoric that diffuses and obfuscates its meaning and weakens its political appeal.

Neither does Allen's claim that both nature and society are constructed help the case very much. Theorists such as Foucault (1966), Latour (1994), Haraway (1991) and Knorr-Cetina (1996) have produced convincing reasons for thinking that the distinction between nature and society often precludes learning and can be utilized to reinforce power distributions. The view that a market economy is somehow "natural" is only one of the most egregious abuses of the nature/society fallacy. Yet some nature/societies are clearly more just than others, once some criterion of justice has been proposed, and some nature/societies are probably more sustainable than others. Recognizing that we live (and have always lived) not in nature and society but in a nature/society such that the reasons for distinguishing the two are always pragmatic and value-laden does not suddenly produce a research agenda for social sustainability. It may still turn out that resource sufficiency or ecological sustainability are precisely the concepts that we need, not because they are "natural," but for pragmatic and value-based reasons. Left-leaning social sustainability needs an account of why its favored values might be thought to contribute to the regeneration of food systems, just as Dahlberg suggests, but in order to get that account, its advocates must frame a research agenda that shows how values of any sort (even profitability) might be relevant to the regeneration of a food system.

Moral Economy

In stating that agriculture must distribute goods fairly and have a participatory decision process, Douglass and Allen have done nothing more than state moral preconditions for sustainable practices. As I have argued, this provides little insight into sustainability as such. It is as if either resource sufficiency or ecological components provide the conceptual content of sustainability, but that any proposal for sustainable development or sustainable agriculture must meet the independent moral criteria favored by Douglass, Thrupp, Regan and the others in their camp. This is a perfectly coherent moral position, but it is not a third paradigm and it does not specify any research agenda for social sustainability. There may be a need for research on liberal social institutions, but that need exists apart from any conceptualization of sustainability.

A more promising start can be given to the leftist view by placing it within the context of research on moral economy. E.P. Thompson's paper "The Moral Economy of the English Crowd in the 18th Century" laid the foundations for the last three decades of work. Thompson attempted to explain bread riots in rural English villages as a protest against the emergence of market structures for distributing grain. Put simply, English peasants and villagers thought themselves to have an entitlement to purchase grain and bread from the fields of nearby farms in their district. This perceived entitlement was challenged by farmers who were beginning to use improved roads and to seek the best markets for grain. Although villagers recognized that farmers were entitled to compensation for grain, they rejected the farmer's right to exchange with any willing buyer, and hence felt that the larger regional markets caused an illegitimate and extortionary price increase for bread and grain. (Thompson, 1971.)

Thompson used the term "moral economy" to describe the system of rights, privileges, norms, and expectations that organize—or at least frame—relationships of production, distribution, and exchange in small village societies. Moral economy provides a structure of rules for producing and consuming subsistence commodities, especially food and especially in times of resource scarcity and stress. To a large extent, moral economy is replaced by political economy as central states take power over the roads, currency and other elements of infrastructure that are necessary for regional and national markets. Political economy also frames private transactions within a structure of rights and norms but does so in a manner that is formally institutionalized by public laws and regulations. Thompson believed that political economists (meaning Adam Smith) would be unlikely to endorse the implicit system of rights and rules that comprise moral economy. First, moral economy limits the power and authority of the central state. Moreover, trade is hidden within transactions so private that they easily escape the tax assessor's notice, and finally such private rules discourage the expansion of production and the growth of markets that should, in Smith's theory, make everyone better off.

Thompson's idea was expanded considerably by James Scott in *The Moral Economy of the Peasant*. Scott's ethnographic research in Asia found a vibrant moral economy among peasant farmers. One notable feature was a practice of choosing production methods that minimized the risk of a crop failure, despite clear knowledge that alternative methods were optional from a profit-maximizing perspective. Scott found in effect that peasants organize their society according to the "maximum" rule enshrined by John Rawls (1971) in his "difference principle:" choose the social structure that has the best worst case. Neither Thompson, Scott, nor Rawls wrote about sustainability, but it is only a short step away. The forms of moral economy described by Thompson and Scott are *satisficing* systems of social organization; they aim to maximize the chance of merely sustaining the society, rather that achieving an optimal level of social well-being.

The implication that one might draw is that the historical transition from satisficing to optimizing social institutions creates a situation where people are constantly placed at risk. Thompson's emphasis on the period of transition from feudal agrarian societies to industrial capitalism is certainly consistent with such a view. Yet this view of social sustainability still leaves several key questions unanswered. It is plausible to assume that any system of agriculture that remains stable over a period of centuries is sustainable. We will find few examples of human practices that are recognizably stable for longer periods (Dickson, 1995). So the long-lived precapitalist and peasant systems of food production and distribution provide a good reference point for sustainability. Furthermore, the changes that take place in these systems through the transition to market economy and industrial capitalism are also relevant to use of injustice or inequality as a reference point for sustainability. Nevertheless, the fact that *people* are placed at risk in a system for food production and distribution is not in itself evidence that the *system* is unsustainable. To put it bluntly, the system might simply "consume" a certain number of people—the hungry, the marginalized, and low-wage workers (or slaves)—but so long as the "breeding stock" for these human inputs to the system is maintained, the "flow" of people needed for the production process can simply be "harvested," just as fish are harvested from a sustainable fishery. An agricultural social system that treats the people who suffer and die through its machinations as a harvestable flow is morally repugnant, but is it unsustainable? Can't such systems continue indefinitely?

Moral Economy and Social Sustainability

The moral economy of Thompson and Scott is thus not in itself a key to sustainable society. One conceptual problem with Thompson's moral economy in particular is that it is portrayed as static: a system of rights, rules, and expectations. But clearly such systems are susceptible to change, and as such, we can ask how systems of moral economy either remain stable or evolve over time. Scott's recent work, beginning with *Weapons of the Weak,* and continued in *Domination and the Arts of Resistance,* begins to examine the practices and procedures for the social reproduction of moral economy, or for ensuring that key norms and beliefs are shared continuously and extensively over at least some expanse of space and time. (Scott, 1990) Two of Scott's points are especially relevant. First, he describes the constant testing of rights and constraints through minor affronts and argues that this behavior is as crucial to the social reproduction of moral economy as is the verbal telling and retelling of rights and privileges through official documents and informal channels. The constant give and take permits a gradual evolution in the terms of moral economy, especially with regard to claims that relatively powerful people make upon the less powerful, and vice versa. Second, he shows that informal channels that can be hidden from view are especially important for the reproduction of moral economy among the weak. Furthermore, these informal communication networks can consist of highly oblique "texts," including folktales, fables, theater and carnival. The vagueness of these texts permits multiple interpretations, and hence spawns a hidden discourse of testing and enforcement even within relatively powerless groups.

I want to propose that collectively these verbal and non-verbal procedures of reproduction, testing and revision constitute the practical moral discourse that aims to reform moral economy in Thompson's sense. The potential for revision and transformation of rights, privileges and constraints (that is, the potential for moral discourse) implies an ethical dimension: verbal and non-verbal exchange aims not at what the moral economy *is*, but at what it *ought* to be. This is still a long way from what we ordinarily call moral philosophy, for practical moral discourse as I have described it implies no systematization of moral claims. It does not require that agents in a moral exchange apply standards of consistency to their claims or behavior. Furthermore, many of the actions or strategies described by Scott entail or threaten violence and seek nothing more than individual or group interest. Nevertheless the ensemble affronts, tit-for-tat retributions, and verbal disputes reproduce a moral order, which however far from a philosophical ideal, all the same permits certain sorts of action without fear of retribution and constrains other actions that might be otherwise attempted.

The linguistic and non-linguistic practices of reproduction and revision embody a second order that refers to the first-order structure of rights, privileges and constraints. I have called this second order of moral economy "moral discourse," but it does not meet the political or ethical theorists' notion of morality. The conflict and negotiation dimension of moral discourse reproduces moral economy at the first or-

Hierarchy of Normativity

First-Order Moral Economy	Structure of rights, privileges and constraints
Second-Order Moral Economy	Moral discourse or negotiation of structure
Third-Order Moral Economy	Political and ethical theory or systematic idealizations of practical moral discourse

der: the structure of rights, privileges and constraints. But the strategic dimension of moral discourse, the element of challenging opponents claims and imposing costs on their attempts to make claims produces a normative dimension to moral discourse that stipulates reflexively what the first-order structure should be. Conventional political and ethical theory can be interpreted as an extremely purified version of moral discourse, purged of violent and coercive dimensions.

This notion of practical moral discourse may come perilously close to what many of us mean when we use the word "values." E. P. Thompson worries that "if values, on their own, make a moral *economy* then we will be turning up moral economies everywhere." (Thompson, 1993, p. 339.) This not only carries the potential of making moral economy into a conceptually empty phrase, it converts Scott's work into a simple restatement of the basic question of social psychology, namely, "How are social norms transmitted from person to person and reproduced over time?" In one sense, I *am* simply saying that social sustainability does not become a meaningful research paradigm until social psychology is placed at the center, but I also believe that moral economy is a new and promising way to think about this old problem in sociology. Although it certainly is possible to spell out systems of moral values, meaning a logically consistent account of moral concepts along with rules for determining the relationships among concepts and for generating prescriptions for action, it is doubtful that many individuals actually possess or utilize a "value system," in that sense. Why would we expect an ordered value system to emerge at the community level? The concept of moral discourse, determined as second-order moral economy, places both practices and verbal strategies within a context where continuous deployment of any single package of practices and strategies by a single group or individual[1] will eventually be met by opponents who either attempt to revise the package or impose an alternative. Revision rearranges the implications of practical moral discourse so that it serves a different configuration of rights, privileges and constraints (e.g., a different first-order moral economy). Moral discourse, or second-order moral economy, at least puts the *economy* back by combining linguistic strategies for revising

1. My package is a less grandiose version of *ideology*, including the coercive practices that enforce it. However, being less grandiose is important! There is no reason to think that people employing a package of practices and arguments in a given time or place are necessarily committed to or captured by the kind of totalizing or systematic belief in an ideal economic, political or moral order implied by the term "ideology." I will continue to use the more modest idea of moral discourse.

first-order moral economy with the affronts, retaliation and threats that make it costly either to propose, defend or reject a moral claim. Such practical moral discourse need not be confined to a special corner of social behavior that can be understood as a "holdout" against capitalism. In fact, practical moral discourse is the medium for institutional innovation at the local level. It is what some have called micropolitics.

As I have defined it, practical moral discourse stands between the structure of rights, privileges and constraints (perceived or actual) that are the object of E. P. Thompson's and Scott's early work, and political or ethical theory, which exists primarily in splendid isolation from actual political conflict. As currently practiced, political and ethical theory operates at a third order of normativity, providing an account of what moral discourse *should* be. Conventional ethical and political theory provides a systematic set of rules and concepts for making normative arguments about whether a given structure or pattern of conduct is legitimate, just or morally good. However, in excluding the sometimes violent, sometimes cynical, and always chaotic "negotiation" that goes on when individuals and groups both reproduce and revise the structure of norms—the moral economy—political and ethical theorists leave a gap between their own discourse and practice. Scott's later work can be interpreted as an attempt to theorize what happens in that gap. He describes an arena where the structure of rights, privileges and constraints is reproduced and revised; reproduction and revision are fully economic processes in that they have costs, risks and benefits. It is Scott's appreciation of the economic dimension to the maintenance of norms that provides an opening for rethinking this process as a problem in sustainability.

When the middle level is ignored, we can conceive of political and ethical theory as performing the critique of social structure, but this is only plausible to the extent that we think of social structure itself being reproduced entirely by formal mechanisms such as law, education and government. Many of E. P. Thompson's readers may have thought that it was industrial capitalism's ability to reproduce itself through these formal state-sponsored mechanisms that marks the difference between moral and political economy, but this picture of social transformation has some evident flaws. For one thing, it implies that the vast majority of the populace participate in moral discourse vicariously, through the process of elections or as passive recipients of moral and political ideology. One should question whether norms are ever reproduced this way at all. What seems more likely is that a new kind of hidden discourse will begin to emerge, one that takes place out of view from the public, official state-sponsored discourse. This discourse will have a very different shape and texture than that of Scott's peasants, for in open societies being "out of view" may not literally mean "hidden," and the testing and conflict that shapes this discourse in stratified class societies will also be very different.

Yet it is arguably only at the middle level that we can begin to reformat questions about social regeneration. Social sustainability depends on whether practical moral discourse reliably and authentically reproduces and appropriately modifies the structure of rights, privileges and constraints that gives a society its distinctive identity and culture. Perhaps even more crucially, second-order moral discourse links personal interests and felt personal loyalties to the moral language of duty, responsibility, rights and accountability. This means that practical moral discourse is a crucial element of *effective* norms, norms that function *as* norms, rather than simply as codifications of ideology and state power.

The reason that I have shunned the development of a recognizable political or ethical theory in this book is not that I think the idealizations associated with such theories are inappropriate. Indeed, each of us must contemplate our world in somewhat idealized terms in order to have any moral anchor at all. Yet having a sustainable agriculture (or indeed a sustainable society) demands some linkage between our actual policies (and the battles to shape them) and the idealizations that are the proper subject of political and moral theory. We need this link because theory does not contribute to sustainability unless it can inform practice broadly enough to affect the obvious biological performance indicators of the food system (e.g., fertility, productivity, etc.). And theory cannot have this informative aspect if it is confined to the academy. We need, in short, practical moral discourse. Practical moral discourse should be the academy's outreach to the broader society, (which is why I have restructured the Extension leg of the land-grant mission so dramatically). Practical moral discourse should inform research, and preparation for practical moral discourse should be the first objective of our teaching.

Moral Discourse and Sustainable Agriculture

Moral discourse is essential to agriculture in so far as people's willingness to recognize morally valid rights, privileges and constraints shapes agricultural practice, and agriculture is essential to moral discourse to the extent that practices for producing and consuming food are sources of conflict, interest and loyalty. Put another way, we may ask if it will ever be possible to have a truly sustainable agriculture in a world where farmers and agriculture's key decision makers are entirely cut off from the experience of nature, the approval of community, or the claims of the hungry. We may also ask whether a sustainable society is possible in a world where food is provided in devicelike fashion, and where the problem of food availability ceases to pervade the practical moral discourse of the common person. It seems likely that people living in such a world would lack the practical moral vocabulary to even form the idea that agriculture might have unique responsibilities to (and dependency on) nature, the community and the hungry and, lacking these ideas, might fail to see some of the special considerations awarded to agriculture as components of reciprocity that have emerged out of practical moral discourse.

Practical moral discourse ends to the extent that the daily lives of people cease to be informed by a structure of rights and expectations born of personal loyalty and conflict. As conflicts tend to be mediated by the state, practical moral discourse gives way to politics. There is no sharp distinction between the two, but at the more politicized extreme, one can question whether conflict and negotiation any longer retain the capacity to reproduce the vocabulary of rights, responsibilities, virtues, vices, and legitimate expectations. People who lack an experiential basis for these concepts are left with nothing but market incentives to guide their practice and may construct a politics that is utterly uninformed by ethics and history. There are, thus, reasons to think that a system of agricultural production which places key agricultural decision makers in a direct dependence on local ecological processes and on local community support may indeed be more sustainable. It is impossible to elaborate on how emerging industrialized systems may fail to do this within the confines of a concluding essay, so one suggestive example that stresses a philosophical point must suffice.

Research on resource sufficiency and ecological sustainability has proved capable of specifying the human behavior needed to reach goals in at least a few instances. As stated above, fisheries management can set flow levels so that fish will be regenerated and fisheries are sustainable. Many of the scientists who have established these targets assume that once people have been told what course of action is needed to reach an agreed upon goal, simple rationality dictates that they will pursue it. Yet, of course, this seldom happens in practice. One reason is that strategic considerations can make individual rationality diverge from the cooperative behavior needed to conserve resources (Lee, 1993.) A more philosophical reason may be that the utilitarian morality implied by this means-end model of environmental management is fundamentally flawed. This simple picture of fisheries management presumes that the morally good action is the one that reliably produces the morally good outcome. It implicitly presumes that people conceive of morality as a problem of choosing the means to bring about morally justified consequences. Difficult work is presumed to lie in evaluating the cost/benefit trade-offs implied by any course of action, but once that work is done, morality consists simply in performing the actions that bring about the best consequences.

Predictive capacity is a cornerstone of experimental science. Hypotheses that do not engender testable predictions are discarded as speculative and unresearchable. Successful science establishes generalizations that can be utilized to predict real-world behavior of physical and biological systems. Applied science goes one step further, building these generalizations into technological control systems. The combined ability to predict and control gives one the ability to predict future states of the world, each of which is contingent upon a given manipulation of the control system today. As argued in Chapters 1 and 2, utilitarianism comes naturally to someone steeped in the methods and norms of applied science. Put another way, applied science seems to empower the utilitarian moralist by providing a technological means for defining discrete options and equipping it with science-based prediction of the consequences that each option would produce if chosen.

The first lesson on the road to sustainable agriculture is that this empowerment is illusory. First, as argued in Chapters 1 and 2, the empowered utilitarian is quite likely to have the same blind spots as a utilitarian theorist. Second, as argued in Chapter 3, in providing the technological means for pursuing one set of options, applied science can unwittingly foreclose others. Ultimately, the most limiting foreclosure of options occurs in the attenuation of moral discourse: one cannot think of ethical issues *except* in utilitarian or consequentialist terms. The BST debate, visited at intervals throughout the book, pits a science and industry coalition against a rabble of other interests. Yet by seeing the risk issues and the animal health issues purely in consequence-predicting terms, and in failing to recognize the validity of discourse framed in terms of rights, consent, participation or autonomy, the science/industry coalition alienates and frightens a broader public. They win the regulatory battle (BST is approved for use), but at what cost to the bonds of public trust in science?

Ironically, the consequence-predicting character of their wisdom imprisons applied scientists and utilitarian moralists. Although they can see that a given social structure or technological regime will generate dysfunctional outcomes, they lack a language that equips them to participate fully in society's moral discourse. The second lesson on the road to sustainable agriculture is that public trust in science is a product of moral discourse, of second-order moral economy. It comes from the experience of winning just battles with the aid of science, and from recognizing injustice (or error in oneself) when science seems to line up with the wrong side. This trust can withstand a deployment of science in an unpopular cause, but it withers when the scientist *qua scientist* seems constitutionally blinded to an entire spectrum of moral issues. The land-grant system's seduction by utilitarian morality has, to all appearances, done just that. As noted in the Introduction, the transition from agrarian to trading societies spawned a revolution in moral vocabulary, but the revolution is incomplete. The language of rights and consequences has never fully displaced the language of loyalty, integrity and virtue. Moral suasion couched entirely in terms of predicted consequences fails to persuade.

However, things may be worse yet. The agrarian philosophy of agriculture held that farming was special. We can re-articulate the lessons of Jefferson, Emerson and Berry in the language developed in this concluding chapter. The food system of a society establishes the material basis for moral economy, that is, for negotiating some of the most basic and ineliminable problems of human existence. We may call this dimension of the material basis for society its *agricultural constitution*. We may ask how the moral discourse that ensues from a society's agricultural constitution equips its citizens with a vision of the common good, a capacity to act in pursuit of justice and a conception of nature. When Jefferson and Emerson asked this question, they pointed to the way that farming resolved a potential conflict between personal and public interest, or to the way that solving subsistence problems within nature's constraints led to the formation of a virtuous character.

When I ask this question with respect to North America at the third millennium I am not comforted. Where one might find a vision of the common good I find only the endless multiplication of purely individualized choices. There are miles of aisles, and each is brimming with consumer choices, but there is no vision of the common interest in *producing* food. Justice, too, is an attenuated moral discourse, addressed only in terms of helping the poor join the orgy of consumption underway at the grocery stores and fast-food outlets that line America's boulevards. There is no mention of justice in production, no thought that as we produce our food we might be reproducing an agricultural constitution that in some measure allows us to function as citizens (or not).

Most troubling, however, is that I find no conception of nature or human character emerging from our moral food economy at all. Neither food consumption (shopping and dining out) nor the occasional public debate over food stamps, subsidies, pesticides or the FDA yields any conception of humanity's dependence on land and water. Any recognition of a responsibility to conserve those resources wisely arises from other quarters, far from the discourse in which we negotiate our access to food. If the sustainability of our republic or our bodies depends in any measure on our agricultural constitution's capacity to build character and to generate enlightening moral discourse, we are in deep, deep trouble. At times it seems our best hope is to believe that Jefferson must simply have been wrong, and that the wellsprings of citizenship and stewardship of nature simply lie elsewhere.

Yet my better instincts say that while we need not be a nation of farmers, we ought not to become a nation as blinded to agrarian moral discourse and the negotiation of our agricultural constitution as we presently appear to be. The land-grant mission for sustainable agriculture is to frame research questions in light of a broader conception of agriculture's role. Social science and humanities especially must be called upon to examine how and whether material food systems affect our broader society's capacity to envision a nature/society in sustainable terms. The land-grant mission for sustainable agriculture is to teach explicitly what Emerson thought would be the natural birthright of those who grew up working in the presence of nature: a recognition that work is imbued with value. Agricultural ethics must link self-reliance to citizenship through the medium of nature, and we sustain our society when we teach this link to our children. Finally, the land-grant mission for sustainable agriculture is to reinvigorate our policy debates on food systems by infusing them with a richer moral vocabulary. This is not to say that we should become moralistic, but a policy dialog conducted solely in terms of costs and benefits, of choices and their consequences, will suffice no longer. We must diversify and deepen our capacity for saying what is important to us and why, and for hearing what is important to others (and why). Perhaps then we will be able to act collectively and hopefully sustainably.

References

Aiken, William and Hugh LaFollette. 1995. *World Hunger and Moral Obligation,* 2d edition. Englewood Cliffs, N.J.: Prentice-Hall.

Alexander, Martin. 1985. "Ecological consequences: Reducing the uncertainties," *Issues in Science and Technology* 13:57-68.

Allen, Patricia, ed. 1993. *Food for the Future: Conditions and Contradictions of Sustainability.* John Wiley & Sons, Inc., New York.

Allen, Patricia and Carolyn E. Sachs. 1992. "The poverty of sustainability: An analysis of current positions." *Agriculture and Human Values.* Fall, 94:29-35.

Altieri, Miguel. 1987. Agroecology. The Scientific Basis of Affirmative Agriculture. Boulder, Colo.: Westview Press.

Aristotle. 1957. *Aristotle's Politics and Poetics.* B. Jowett and T. Twining, trans. New York; The Viking Press.

Aristotle. 1962. *Nichomachean Ethics.* M. Oswald, Trans. Indianapolis, Ind.: The Liberal Arts Press.

Attfield, Robin. 1983. *The Ethics of Environmental Concern.* New York: Columbia University Press.

Austin, Richard C. 1990. *Reclaiming America: Restoring Nature to Culture.* Abingdon, VA: Creekside Press.

Baier, Annette. 1986. "Poisoning the wells." In *Values at Risk,* Douglas E. MacLean, ed., Totowa, NJ: Rowman and Allanheld, pp. 49-74.

Batie, Sandra S. 1984. "Soil conservation policy for the future." *The farm and food system in transition: Emerging policy issues,* No. 23. Lansing: Cooperative State Extension Service, Michigan State University.

Bellah, Robert, Robert Madsen, William M. Sullivan, Ann Swidler and Steven M. Tipton. 1985. *Habits of the Heart: Individualism and Commitment in American Life.* New York: Harper & Row.

Bentham, Jeremy. 1789. *The Principles of Morals and Legislation* (republished 1948). New York: Hafner Publishing Co.

Berry, Wendell. 1972. "Discipline and hope." In *A Continuous Harmony.* New York: Harcourt Brace Jovanovich.

Berry, Wendell. 1977. *The Unsettling of America.* San Francisco: Sierra Club Books.

Berry, Wendell. 1981. *The Gift of Good Land: Further Essays, Cultural and Agricultural.* San Francisco: North Point Press.

Berry, Wendell. 1983. *A Place on Earth.* San Francisco: North Point Press.

Berry, Wendell. 1987. *Sabbaths.* San Francisco: North Point Press.

Betts, E.M., ed. 1953. *Thomas Jefferson's Farm Book.* Princeton, NJ: Princeton University Press.

Bishop, K. 1987. "California U. told to change research to aid small farms." *The New York Times.* November 19, 1987:13A.

Blatz, C. Ed. 1991. *Ethics and Agriculture.* Moscow: University of Idaho Press.

Bonnen, J. T., and Browne, W. 1989. "Why is agricultural policy so difficult to reform?" In C. S. Kramer, Ed., *The political economy of U.S. agriculture* pp. 7-15. Washington, D.C.: Resources for the Future.

Brandt, Jon A., and Ben C. French. 1983. "Mechanical harvesting and the California tomato industry." *American Journal of Agricultural Economics* 65:265-272.

Breimyer, H.F. 1965. *Individual Freedom and the Economic Organization of Agriculture.* Urbana, Ill: University of Illinois Press.

Bromley, Daniel W. 1989. *Economic Interests and Institutions: The Conceptual Foundations of Public Policy.* New York: Basil Blackwell.

Browne, W., J. Skees, L. Swanson, P. Thompson, and L. Unnevehr. 1992. *Sacred Cows and Hot Potatoes: Agrarian Myths and Policy Realities,* Boulder, CO: Westview Press.

Browne, William P. 1987. "Bovine Growth Hormone and the Politics of Uncertainty: Fear and Loathing in a Transitional Agriculture." *Agriculture and Human Values* IV 1:75-80.

Buchanan, James. 1987. *Economics: Between Predictive Science and Moral Philosophy.* College Station: Texas A&M University Press.

Burkhardt, Jeffery. 1991. "Ethics and Technical Change: The Case of BST." Center for Biotechnology Policy and Ethics Discussion Paper CBPE 91-2, Texas A&M University, College Station, Texas.

Burton, Jeanne L. and Brian W. McBride. 1989. "Recombinant, Bovine Somatotropin rBST: Is There a Limit for Biotechnology in Applied Animal Agriculture?" *Journal of Agricultural Ethics* 22:129-160.

Busch, L., W. Lacy, J. Burkhardt and Laura Lacy. 1991. *Plants, Power and Profit.* Cambridge, MA: Basil Blackwell.

Busch, Lawrence and William Lacy. 1983. *Science, Agriculture and the Politics of Research.* Boulder, CO: Westview Press.

Buttel, F.H. 1986. "Agricultural Research and Farm Structural Change: Bovine Growth Hormone and Beyond," *Agriculture and Human Values* 3:88-98.

Callicott, J. Baird. 1990. The Metaphysical Transition in Farming: From "The Newtonian-Mechanical in the Eltonian-Ecological." *Journal of Agricultural Ethics* 3:113–146.

Carson, R. 1962. *Silent Spring.* Boston: Houghton Mifflin.

Cochrane, W. 1979. *The Development of American Agriculture: A Historical Analysis.* Minneapolis: Minnesota Press.

Cochrane, W. 1985. "The need to rethink agricultural policy in general and perform some radical surgery on commodity programs in particular." *American Journal of Agricultural Economics,* 67, 1002-1009.

Comstock, Gary. 1987. *Is There a Moral Obligation to Save the Family Farm?* Ames: Iowa State University Press.

Comstock, Gary. 1988. "The Case Against bGH," *Agriculture and Human Values* 53:36-52.

Copp, David. 1985. "Morality, reason and management science: The rationale of cost benefit analysis." *Social Philosophy and Policy* 2:129-151.

Corrington, Robert S. 1990. "Emerson and the Agricultural Midworld." *Agriculture and Human Values* 71:20-26.

Covello, Vincent T., Peter M. Sandman, and Paul Slovic. 1991. "Guidelines for Communicating Information About Chemical Risks Effectively and Responsibly." In *Acceptable Evidence: Science and Values in Risk Management,* Deborah G. Mayo and Rachelle D. Hollander eds., New York: Oxford University Press.

Cross, Frank, B. 1992. "The Risk of Reliance on Perceived Risk." In *Risk: Issues in Health and Safety,* 31: 59-70.

Dahlberg, K. 1993. "Regenerative Food Systems: Broadening the Scope and Agenda of Sustainability." In *Food for the Future.* Allen, Patricia, ed. New York: John Wiley and Sons, Inc., 75-102.

Dahlberg, K., ed. 1986. *New Directive for Agricultural Research: Neglected Dimensions and Emerging Alternatives.* Totawa, N.J.: Rowmand and Allenheld.

Danbom, D.B. 1986. "Publicly sponsored agricultural research in the United States from an historical perspective," *New Directions for Agriculture and Agricultural Research: Neglected Dimensions and Emerging Alternatives.* Dahlberg, K.A., ed.: Totowa, N.J.: Rowman and Allanheld, 142-162.

DeWalt, Billie R. 1988. "Halfway There: Social Science in Agricultural Development and the Social Science of Agricultural Development." *Human Organization* 47:343–353.

Dewey, John. 1939. *Freedom and Culture.* New York: Paragon Books.

Douglass, Gordon K., ed. 1984. *Agricultural Sustainability in a Changing World Order.* Boulder, Colo.: Westview Press.

Doyle, Jack. 1985. *Altered Harvest.* New York: Viking/Penguin Books.

Doyle, Michael P., and Elmer H. Marth. 1991. "Food Safety Issues in Biotechnology." *Agriculture Biotechnology: Issues and Choices.* West Lafayette, IN: Purdue University Agricultural Experiment Station pp. 55-66.

Dundon, S. 1986. "The moral factor in innovative research." In *The Agricultural Scientific Enterprise,* ed. L. Busch and W. Lacy. Boulder, Colo.: Westview Press.

Dworkin, Ronald. 1977. *Taking Rights Seriously.* Cambridge, Mass.: Harvard University Press.

Emerson, Ralph Waldo. 1870, reprinted 1904. "Farming," from *"Society and Solitude."* In *The Complete Works of Ralph Waldo Emerson.* Concord Edition Vol. 7. Boston: Houghton Mifflin.

Emerson, Ralph Waldo. 1904. *Society and Solitude.* Boston: Houghton Mifflin.

Emerson, Ralph Waldo. 1965. *Selected Writings of Ralph Waldo Emerson.* Gilman, William H. ed. Ontario: The New American Library.

Feinberg, Joel. 1970. "The Nature and Value of Rights." In *The Journal of Value Inquiry,* 44 winter: 243-256.

Fischhoff, Baruch and co-authors. 1981. *Acceptable Risk.* New York: Cambridge University Press.

Flora, C.B. and M. Tomecek, eds. 1986. *Farming Systems Research and Extension:*

Management and Methodology. Manhattan: Kansas State University Press.

Foucault, Michel. 1966, translated 1970. *The Order of Things.* New York: Random House.

Fox, M.W. 1986. *Agricide: The Hidden Crisis that Affects Us All.* Schocken, New York.

Frey, Ronald G. 1983. *Interests and Rights.* Oxford: Oxford University Press.

Frey, Ronald G. 1995. "The Ethics of Using Animals for Human Benefit," in *Issues in Agricultural Bioethics.* T.B. Mepham, G.A. Tucker, and J. Wisemen, eds. Nottingham, U.K.: University of Nottingham Press.

Gendel, S., D. Kline, M. Warren and F. Yates, eds. 1990. *Agricultural Bioethics.* Ames: Iowa State University Press.

Gewirth, Alan. 1982. *Human Rights: Essays on Justification and Applications.* University of Chicago Press: Chicago.

Gibbard, Allan. 1986. "Risk and Value." In *Values at Risk,* Douglas E. MacLean, ed., Totowa, N.J.: Rowman and Allanheld, pp. 94-112.

Giddens, Anthony. 1984. *The Constitution of Society.* Berkeley and Los Angeles: University of California Press.

Giere, Ronald. 1991. "Knowledge, Values, and Technological Decisions: A Decision Theoretic Approach." In *Acceptable Evidence: Science and Values in Risk Management,* Deborah G. Mayo and Rachelle D. Hollander, eds., New York: Oxford University Press, pp. 183-203.

Glickman, Theodore S., and Michael Gough eds. 1990. *Readings in Risk.* Washington, D.C.: Resources for the Future.

Goldstone, Jack A. 1991. *Revolution and Rebellion in the Early Modern World.* Berkeley: University of California Press.

Graham, John D., Laura C. Green, and Marc J. Roberts. 1988. *In Search of Safety: Chemicals and Cancer Risk,* Cambridge, Mass.: Harvard University Press.

Griffen, James. 1982. "Modern Utilitarianism," *Revue Internationale de Philosophie* 141:131-175.

Hadwiger, Don F. 1982. *The Politics of Agricultural Research.* Lincoln: University of Nebraska Press.

Hallberg, M.C. 1992. *Bovine Somatotropin and Emerging Issues: An Assessment.* Boulder, Colo.: Westview Press.

Hamilton, Alexander, John Jay and James Madison. 1964. *The Federalist Papers.* New York: Washington Square Press.

Hanson, Michael. 1991. "Consumer Concerns: Give Us the Data." In *Agricultural Biotechnology at the Crossroads: Biological, Social and Institutional Concerns,* Ithaca, N.Y.: National Agricultural Biotechnology Council, pp. 169-178.

Haraway, Donna. 1991. *Simians, Cyborgs and Women.* New York: Routledge Publishing Company.

Hardin, G. 1968. "The tragedy of the commons." *Science,* 1243-1248.

Harlander, Susan K., James N. BeMiller, and Larry Steenson. 1991. "Impact of Biotechnology on Food and Nonfood Uses of Agricultural Products." *Agriculture Biotechnology: Issues and Choices.* West Lafayette, Ind.: Purdue University Agricultural Experiment Station. pp. 41-54.

Hess, Charles E. 1984. "Freedom of inquiry-an endangered species." Presentation to the Division of Agriculture, National Association of State and Land Grant Universities, Denver, Colo., 13 November 1984.

Heyboer, M. 1995. The Normative Implications of the Configuration of the Applied Sciences. In *New Directions in the Philosophy of Technology.* J. Pitt, ed. Dordrecht, The Netherlands: Kluwer Academic Publishers, 153-157.

Hightower, Jim. 1975a. The Case for the Family Farm. In *Food for People, Not for Profit.* Lerza, C., and Jacobson, M., eds. New York: Ballantine.

Hightower, Jim. 1975b. *Eat Your Heart Out.* New York: Crown Publishers.

Hightower, Jim. 1978. *Hard Tomatoes, Hard Times.* Cambridge, Mass.: Schenkman Publishing Company.

Hollander, Rachelle D. 1991. "Expert Claims and Social Decisions: Science, Politics, and Responsibility." In *Acceptable Evidence: Science and Values in Risk Management,* Deborah G. Mayo and Rachelle D. Hollander, eds., New York: Oxford University Press, pp. 160-173.

Hopkins, D. Douglas, Rebecca J. Goldberg and Steven A. Hirsch. 1991. *A Mutable Feast: Assuring Food Safety in the Era of Genetic Engineering,* New York: Environmental Defense Food.

Jackson, Wes. 1980. *New Roots for Agriculture.* San Francisco: North Point Press.

Jasanoff, Sheila. 1991. "Acceptable Evidence in a Pluralistic Society." In *Acceptable Evidence: Science and Values in Risk Management,* Deborah G. Mayo and Rachelle D. Hollander, eds., New York: Oxford University Press, pp. 29-30.

Jefferson, T. 1984a. Letter to Jay. In M.D. Peterson, ed., *Writings* pp. 818-820. New York: Literary Classics of the United States.

Jefferson, T. 1984b. Notes on the state of Virginia. In M. D. Peterson, ed., *Writings* pp. 281-293. New York: Literary Classics of the United States.

Jefferson, Thomas. 1984c. *Writings.* Merrill D. Peterson, ed. New York: Literary Classics of the United States.

Johnson, Branden B. and Vincent T. Covello eds. 1987. *The Social and Cultural Construction of Risk: Technology, Risk, and Society.* Norwell, MA: Kluwer Academic Publishers.

Johnson, Glenn L. 1982. "Agro-Ethics: Extension, Research, and Teaching." *Southern Journal of Agricultural Economics,* July pp. 1-10.

Johnson, Glenn L. 1984a. "Academia needs a new covenant for serving agriculture." Mississippi State Agricultural and Forestry Experiment Station Special Publication.

Johnson, Glenn L. 1984b. "Ethics, Economics, Energy, and Food Conversion Systems." In *Food and Energy Resources.* New York: Academic Press.

Johnson, Glenn L., and Paul B. Thompson. 1991. "Ethics and Values Associated with Agricultural Biotechnology." *Agriculture Biotechnology: Issues and Choices.* West Lafayette, IN: Purdue University Agricultural Experiment Station. pp. 121-138.

Johnson, Glenn L., James T. Bonnen, eds. 1991. *Social Science Agricultural Agenda and Strategies.* East Lansing: Michigan State University Press.

Jonas, Hans. 1984. *The Imperative of Responsibility: In Search of an Ethics for the Technological Age.* Chicago, Ill.: University of Chicago Press.

Kaldor, Donald R. 1971. "Social returns to research and the objectives of public research." *Resource Allocation in Agricultural Research,* edited by W. L. Fischer. Minneapolis: University of Minnesota Press.

Kalter, Robert J. 1985. "The new biotech agriculture: Unforeseen consequences." *Issues in Science and Technology* 21: 125-133.

Kant, Immanuel. 1983. *Perpetual Peace and Other Essays.* T. Humphreys, tr. Indianapolis, Ind.: Hackett Publishing Co.

King, Lauriston and Kimberly McGar Stephens. 1992. "The Past and Possible Future of the Animal Rights Movement in the United States," Texas A&M University: Center for Biotechnology Policy and Ethics Discussion Paper CBPE-92-4.

Kirkendall, R.S. 1987. "Up to now: A history of American agriculture from Jefferson to revolution to crisis." *Agriculture and Human Values.* 4(1):4-26.

Kloppenburg, Jack, Jr. 1984. "The social impacts of biogenetic technology in agriculture: past and future." *The Social Consequences and Challenges of New Agricultural Technologies,* edited by G. M. Berardi and C. C. Geisler. Boulder, Colo.: Westview Press, Inc.

Knorr, Bryce. 1991. "Ethics and the American Farmer," *Farm Futures* 191:10-14.

Knorr-Cetina, Karen. 1996. "Epistemics in Society," in *Rural Reconstruction in a Market Economy,* W. Heijman, H. Hetsen and J. Frouws, eds. Wageningen, The Netherlands: Mansholt Studies, 5, pp 55–73.

Kroger, Manfred. 1992. "Food Safety and Product Quality." In *Bovine Somatotropin and Emerging Issues: An Assessment,* Milton, C. Hallberg ed., Boulder, Colo.: Westview Press, pp. 265-270.

Kuchler, Fred, John McClelland, and Susan E. Offutt. 1990. "Regulatory Experience with Food Safety: Social Choice Implications for Recombinant DNA-Derived Animal Growth Hormones." In *Biotechnology: Assessing Social Impacts and Policy Implications,* David J. Webber ed., Westport, Conn.: Greenwood Press, pp. 131-144.

Kuhn, Thomas. 1970. *The Structure of Scientific Revolutions.* 2nd ed. Chicago: University of Chicago Press.

Lacy, William B. and Lawrence Busch. 1991. "The Fourth Criterion: Social and Economic Impacts of Agricultural Biotechnology." In *Agricultural Biotechnology at the Crossroads: Biological, Social and Institutional Concerns, NABC Report 3,* June Fessenden MacDonald, ed., New York: National Agricultural Biotechnology Council.

Lacy, William B., Lawrence Busch, and Laura R. Lacy. 1991. "Public Perceptions of Agricultural Biotechnology." *Agriculture Biotechnology: Issues and Choices.* West Lafayette, Ind.: Purdue University Agricultural Experiment Station, pp. 139-162.

Lanyon, L.E., and D.B. Beegle. 1989. "The role of on-farm nutrient balance assessments in an integrated approach to nutrient management." *Journal of Soil and Water Conservation* 44:164-168.

Lappe, Frances Moore. 1971. *Diet for a Small Planet.* New York: Ballantine Books.

Lappe, Frances Moore. 1985. "The Family Farm: Caught in the Contradictions of American Values." *Agriculture and Human Values* 22:36-43.

Laszlo, Ervin. 1972. *The Systems View of the World.* New York: George Braziller.

Latour, Bruno. 1988. *Science in Action.* Cambridge, Mass.: Harvard University Press.

Latour, Bruno. 1994. *We Have Never Been Modern.* Cambridge, Mass.: Harvard University Press.

Lee, Kai. 1993. *Compass and Gyroscope.* New York: Island Press.

Leopold, Aldo. 1947. *A Sand County Almanac and Sketches Here and There.* Oxford: Oxford University Press.

Lewis, H.W. 1990. *Technological Risk.* New York: W.W. Norton & Company, Inc.

Locke, John. 1690. *Second Treatise of Government,* reprinted 1989, C.B. MacPherson, ed., Indianapolis: Hackett Publishing Co., pp. 23-24.

Lyon, John. 1987. "*Home Economics* by Wendell Berry." *Agriculture and Human Values* 4(2-3): 118–121.

Machan, Tibor. 1984. "Pollution and political theory." In *Earthbound,* edited by T. Regan. New York: Random House.

MacIntyre, Alisdair. 1977. "Utilitarianism and cost-benefit analysis." In *Values in the Electric Power Industry,* edited by K. Sayre. Notre Dame, Ind.: University of Notre Dame Press.

MacIntyre, Alisdair. 1981. *After Virtue: A Study in Moral Theory,* Notre Dame: University of Notre Dame Press.

MacLean, Douglas E. 1990. "Comparing Values in Environmental Policies: Moral Issues and Moral Arguments." In *Valuing Health Risks, Costs, and Benefits for Environmental Decision Making: Report of a Conference.* Washington, D.C.: National Academy Press pp. 83-106.

Madden, J. Patrick. 1984. "Regenerative Agriculture: Beyond organic and sustainable food production." *The Farm and Food System in Transition: Emerging Policy Issues,* No. 33. East Lansing: Cooperative Extension Service, Michigan State University.

Madden, J. Patrick. 1986. "Beyond conventional economics-An examination of the values implicit in the neoclassical economic paradigm as applied to the evaluation of agricultural research." In *New Directions for Agriculture and Agricultural Research,* edited by K. Dahlberg. Totowa, N.J.: Rowman and Allanheld.

Marois, James J., James J. Grieshop, and L.J. Bees Butler. 1991. "Environmental Risks and Benefits of Agricultural Biotechnology." In *Agriculture Biotechnology: Issues and Choices.* West Lafayette, Ind.: Purdue University Agricultural Experiment Station, pp. 67-80.

Martin, Phillip L., and Alan L. Olmstead. 1985. "The agricultural mechanization controversy." *Science* 227:601-606.

Mayo, Deborah G. 1991. "Sociological Versus Metascientific Views of Risk Assessment." In *Acceptable Evidence: Science and Values in Risk Management,* Deborah G. Mayo and Rachelle D. Hollander, eds., New York: Oxford University Press, pp. 249-279.

McDermott, J.J. 1987. *The Culture of Experience: Philosophical Essays in the American Grain.* Prospect Heights, Ill.: Waveland.

Merrill, Richard A. and Michael R. Taylor. 1986. "Saccharin: A Case Study of

Government Regulation of Environmental Carcinogens." In *Agriculture and Human Values*, vols 1 and 2, vol. 3: 33-73.
Meyer, Lois. ill. by Ruth Sanderson. 1983. *The Store-Bought Doll*. Racine, Wisconsin: Western Publishing Co.
Mill, John Stuart. [1861] 1979. *Utilitarianism*, edited by G. Sher. Indianapolis, Ind.: Hackett Publishing Company.
Miller, P. 1956. *Errand into the Wilderness*. Cambridge: Belknap.
Molnar, J.J., K.A. Cummins, and P. Nowak. 1990. Bovine Somatotropin: Biotechnology Product and Social Issue in the U.S. Dairy Industry." *Journal of Dairy Science*. Vol. 73, pp. 3084-93.
Montmarquet, James A. 1989. *The Idea of Agrarianism: From Hunter-Gatherer to Agrarian Radical in Western Culture*. Moscow: University of Idaho Press.
Moore, Barrington. 1966. *Social Origins of Dictatorship and Democracy: Lord and Peasant in the Making of the Modern World*. Boston: Beacon Press.
National Research Council. 1972. *Report of the Committee on Research Advisory to the U.S. Department of Agriculture*. National Academy of Sciences, Washington, D.C.
National Research Council. 1975. *Agricultural Production Research Efficiency*. National Academy of Sciences, Washington, D.C.
Nelkin, Dorothy. 1991. "Living Inventions: Biotechnology and the Public," Texas A&M University: Center for Biotechnology Policy and Ethics Discussion Paper CBPE-91-5.
Norton, Bryan. 1995. "Evaluating Ecosystem States: Two Competing Paradigms," *Ecological Economics*. 14:113-127.
Nozick, Robert. 1974. *Anarchy State and Utopia*, New York: Basic Books.
Parfit, Derek. 1984. *Reasons and Persons*. New York: Oxford University Press.
Railton, Peter. 1990. "Benefit-Costs Analysis as a Source of Information About Welfare." *Valuing Health Risks, Costs, and Benefits for Environmental Decision Making: Report of a Conference*. Washington, D.C.: National Academy Press pp. 55-82.
Rasmussen, Wayne D. 1968. "Advances in American agriculture: The mechanical tomato harvester as a case study." *Technology and Culture* 9:531-543.
Rawls, John. 1971. *A Theory of Justice*. Cambridge, Mass.: Harvard University Press.
Regan, Tom. 1983. *The Case for Animal Rights*. Berkeley and Los Angeles: University of California Press.
Regan, Tom. 1985. "The Case for Animal Rights." In *In Defense of Animals*, P. Singer, ed. New York: Basil Blackwell, pp. 89-107.
Regan, Tom. 1993. "Vegetarianism and Sustainable Agriculture: The Contributions of Moral Philosophy," In *Food for the Future*. Allen, Patricia, ed. New York: John Wiley and Sons, Inc.: 103-121.
Rescher, Nicholas. 1966. *Distributive Justice*. Indianapolis, Ind.: Bobbs-Merrill Publishers.
Rescher, Nicholas. 1983. *Risk: A Philosophical Introduction to the Theory of Risk Evaluation and Management*. Washington, D.C.: University Press of America.
Richgels, Carl E., Samuel J. Barrick, R.H. Foell, and others. 1990. "Sustainable

Agriculture, Perspectives from Industry." *Journal of Soil and Water Conservation.* Jan-Feb., 45:31-3.

Rifkin, Jeremy. 1985. *Declaration of a Heretic.* London: Routledge and Keegan Paul.

Rifkin, Jeremy. 1992. *Beyond Beef The Rise and Fall of the Cattle Culture.* New York: Dutton.

Robbins, J. 1987. *Diet for a New America.* Walpole, N.H.: Stillpoint.

Roberts, Tanya, and Eileen van Ravenswaay. 1989. "The Economics of Safeguarding the U.S. Food Supply," Economic Research Service, Agriculture Information Bulletin Number 566, U.S. Department of Agriculture.

Rollin, Bernard E. 1981. *Animal Rights and Human Morality.* Buffalo, N.Y.: Prometheus Books.

Rollin, Bernard E. 1995. *Farm Animal Welfare: Social, Bioethical, and Research Issues.* Ames: Iowa State University Press.

Rosenberg, C. 1976. *No Other Gods.* Baltimore, MD: Johns Hopkins University Press.

Rosenberg, Nathan and L. E. Birdzell, Jr. 1986. *How the West Grew Rich.* New York: Basic Books.

Rousseau, J. 1984. *A Discourse on Inequality.* M. Cranston, Trans. New York: Penguin. Original work published 1755.

Rousseau, Jean-Jacques. 1984. *The Social Contract.* Maurice Cranston, trans. New York: Penguin.

Rowe, William D. 1977. *An Anatomy of Risk.* New York: John Wiley & Sons.

Russell, Milton. 1990. "The Making of Cruel Choices." In *Valuing Health Risks, Costs, and Benefits for Environmental Decision Making: Report of a Conference,* Washington, D.C.: National Academy Press, pp. 15-22.

Ruttan, Vernon W. 1982. *Agricultural Research Policy.* Minneapolis: University of Minnesota Press.

Sagoff, Mark. 1984. "Ethics and economics and environmental law." *Earthbound,* edited by T. Regan. New York: Random House.

Sagoff, Mark. 1986. "Values and Preferences." *Ethics* 96:301-316.

Sagoff, Mark. 1988. *The Economy of the Earth: Philosophy, Law, and the Environment.* New York: Cambridge University Press.

Sandman, Peter M. 1985. "Getting to Maybe: Some Communications Aspects of Citing Hazardous Waste Facilities," *Seton Hall Legislative Journal,* 9:442-465.

Schaffner, Kenneth F. 1991. "Causing Harm: Epidemiological and Physiological Concepts of Causation." In *Acceptable Evidence: Science and Values in Risk Management,* Deborah G. Mayo and Rachelle D. Hollander, eds., New York: Oxford University Press, pp. 204-217.

Schell, O. 1984. *Modern Meat.* New York: Random House.

Schmid, Allen, James D. Shaffer and Eileen O. van Ravenswaay. 1983. "Community Economics: Predicting Policy Consequences," Department of Agricultural Economics, Michigan State University, East Lansing.

Schmid, Allen. 1987. *Property, Power, and Public Choice: An Inquiry into Law and Economics,* 2d ed. New York: Praeger.

Schmitz, Andrew, and David Seckler. 1970. "Mechanized agriculture and social welfare: The case of the mechanical tomato harvester." *American Journal of Agricultural Economics* 52:569-577.

Schumacher, E. F. 1972. *Small Is Beautiful.* New York: Harper & Row.

Scott, James C. 1976. *The Moral Economy of the Peasant.* New Haven, Connecticut: Yale University Press.

Scott, James C. 1985. *Weapons of the Weak.* New Haven, Connecticut: Yale University Press.

Scott, James C. 1990. *Domination and the Arts of Resistance: The Hidden Transcripts.* New Haven, Connecticut: Yale University Press.

Shapin, Steven and Simon Schaffer. 1985. *Leviathan and the Air-Pump: Hobbes, Boyle, and the Experimental Life.* Princeton, N.J.: Princeton University Press.

Shue, Henry. 1980. *Basic Rights: Subsistence, Affluence, and U.S. Foreign Policy,* Princeton: Princeton University Press.

Simon, Julian L. 1980. "Resources, Population, Environment: An Oversupply of False Bad News," *Science* 208: 1431–1437.

Singer, Peter. 1975. "All Animals Are Equal ... " In *Animal Liberation: A New Ethics For Our Treatment of Animals.* New York: Avon Books.

Singer, Peter. 1979. "Killing Humans and Killing Animals," *Inquiry* 22, pp. 145-56.

Singer, Peter, ed. 1985. *In Defense of Animals.* New York: Harper and Row Publishers.

Singer, Peter and Tom Regan. 1989. *Animal Rights and Human Obligations,* 2d edition. Englewood Cliffs, N.J.: Prentice-Hall. 1989.

Slovic, Paul. 1991. "Beyond Numbers: A Broader Perspective on Risk Perception and Risk Communication." In *Acceptable Evidence: Science and Values in Risk Management,* Deborah G. Mayo and Rachelle D. Hollander eds., New York: Oxford University Press.

Smart, J.J.C., and Bernard Williams. 1973. *Utilitarianism For and Against,* New York: Cambridge University Press.

Smith, Boyd with Barbara Bader. Republished 1990. *The Farm Book.* Boston, Mass.: Houghton Mifflin Company.

Solow, Robert. 1993. "Sustainability: An Economist's Perspective," in *Economics of the Environment: Selected Readings.* Robert and Nancy Dorfman, eds. New York: W.W. Norton and Co., pp. 354–370.

Steinbeck, John. 1979. *In Dubious Battle.* New York: Penguin Books. Original work published 1936.

Steinbeck, John. 1979. *The Grapes of Wrath.* New York: Penguin Books. Original work published 1939.

Stone, Christopher D. 1974. *Should Trees Have Standing? Toward Legal Rights for Natural Objects.* Los Altos, Calif.: William Kaufman.

Sundquist, Burt W., and Joseph J. Molnar. 1991. "Emerging Biotechnology: Impacts on Producers, Related Businesses, and Rural Communities." *Agriculture Biotechnology: Issues and Choices.* West Lafayette, Ind.: Purdue University Agricultural Experiment Station. pp. 23-40.

Taylor, Paul W. 1981. "The Ethics of Respect for Nature," *Environmental Ethics* 3: 197–218.
Thompson, Carol. 1987. *My Big Farm Book.* New York: Platt & Munk Publishers.
Thompson, E.P. 1971. "The Moral Economy of the English Crowd in the Eighteenth Century," *Past and Present* reprinted in Thompson, 1993, pp. 185-258.
Thompson, E.P. 1993. *Customs in Common: Studies in Traditional and Popular Culture.* New York: The New Press.
Thompson, O.E., and F. Scheuring. 1984. "From lug boxes to electronics: A study of California tomato growers and sorting crews, 1977." In *The Social Consequences and Challenges of New Agricultural Technologies,* edited G. M. Berardi and C. C. Geisler. Boulder, Colo.: Westview Press, Inc.
Thompson, Paul B. 1986. "Uncertainty Arguments in Environmental Issues," *Environmental Ethics* 81:59-75.
Thompson, Paul B. 1988a. "Ethics in Agricultural Research." In *Journal of Agricultural Ethics,* Taylor and Francis, 1:11-20.
Thompson, Paul B. 1988b. "The Philosophical Rationale for U.S. Agricultural Policy." In *U.S. Agriculture in a Global Setting: An Agenda for the Future.* Tutwiler, M.A. ed. Washington D.C.: Resources for the Future.
Thompson, Paul B. 1988c. "Ethical issues in agriculture: The need for recognition and reconciliation." *Agriculture and Human Values,* 54: 415.
Thompson, Paul B. 1990a. "Agricultural ethics and economics." *Journal of Agricultural Economics Research,* 42: 3-7.
Thompson, Paul B. 1990b. "Agrarianism and the American Philosophical Tradition." In *Agriculture and Human Values* 71:3-8.
Thompson, Paul B. 1990c. "Risk Subjectivism and Risk Objectivism: When Are Risks Real?" In *Risk: Issues in Health and Safety,* Concord, N.H.: Franklin Pierce Law Center, 11:3-19.
Thompson, Paul B. 1991a. "Constitutional Values and the Costs of American Food." In *Understanding the True Costs of Food: Considerations for a Sustainable Food System.* Greenbelt, Mass.: Institute for Alternative Agriculture, Inc. pp. 64-74.
Thompson, Paul B. 1991b. "Technological Values in the Applied Science Laboratory." In *The Technology of Discovery and The Discovery of Technology,* Joseph C. Pitt and Elena Lugo eds., Virginia: Society for Philosophy and Technology.
Thompson, Paul B. 1992a. "Ethical Issues in BST." In *Bovine Somatotropin and Emerging Issues: An Assessment,* Michael C. Hallberg ed., Boulder, Colo.: Westview Press, pp. 33-50.
Thompson, Paul B. 1992b. "Emphasizing the social sciences and humanities." In *Agriculture and the Undergraduate* pp. 208-221. Washington, DC: National Academy Press.
Thompson, Paul B. 1995a. "Risk and Responsibilities in Modern Agriculture." In *Issues in Agricultural Bioethics,* T.B. Mepham, G.A. Tucker, and J. Wiseman, eds. Nottingham: Nottingham University Press pp. 31-45.

Thompson, Paul B. 1995b. *The Spirit of the Soil: Agriculture and Environmental Ethics,* New York and London: Routledge.

Thompson, Paul B. 1996. "Pragmatism and Policy: The Case of Water," in *Environmental Pragmatism,* E. Katz and A. Light, eds. New York: Routlege Pub. pp. 187–208.

Thompson, Paul B. 1997a. *Food Biotechnology in Ethical Perspective.* London and New York: Blackie Academic for Chapman and Hall.

Thompson, Paul B. 1997b. "Ethics and the Genetic Engineering of Food Animals." *Journal of Agricultural and Environmental Ethics,* 10:1–23.

Thompson, Paul B. and Bill A. Stout, eds. 1991. *Beyond the Large Farm: Ethics and Research Goals for Agriculture.* Boulder, Colo., Westview Press.

Thompson, Paul B., G.L. Ellis, and B.A. Stout. 1991. "Values in the Agricultural Laboratory." In *Beyond the Large Farm,* P.B. Thompson and B.A. Stout, eds., Boulder, Colo.: Westview Press pp. 265-279.

Thompson, Paul B., Robert Matthews, and Eileen van Ravenswaay. 1994. *Ethics Public Policy and Agriculture.* New York: Macmillan.

Thoreau, Henry David. 1965. *Walden and Other Writings of Henry David Thoreau.* B. Atkinson, ed. New York: Random House.

Thrupp, Lori Ann. 1993. "Political Ecology of Sustainable Rural Development: Dynamics of Social and Natural Resource Degradation." In *Food for the Future.* Allen, Patricia, ed. New York: John Wiley and Sons, Inc.:, 47-73.

Tweeten, Luther. 1984. "Food for people and profit: Ethics and capitalism." *The farm and food system in transition: Emerging policy issues,* No. 2. East Lansing: Cooperative Extension Service, Michigan State University.

Van den Belt, Henk. 1995. "Measuring the Environmental Utilization Space: Natural or Social Limits?" Presented at the July 1995 meeting of The International Society for Hermeneutics and Science, 12-15 July, 1995, Leusden, Holland.

Wills, Garry. 1978. *Inventing America: Jefferson's Declaration of Independence.* New York: Vintage Books.

Wills, Garry. 1997. "American Adam," *The New York Review of Books,* 6 March 1997, vol XLIV, no. 4, 30–33.

Wilson, Kathleen, and George E. B. Morren. 1990. *Systems Approaches for Improvement in Agriculture and Natural Resource Management.* New York: Macmillan.

World Commission on Environment and Development. 1987. *Our Common Future.* New York: Oxford University Press.

Index

IOWA STATE UNIVERSITY PRESS
2121 South State Avenue
Ames, Iowa 50014

Orders: 1-800-862-6657
Office: 1-515-292-0140
Fax: 1-515-292-3348

ISBN 0-8138-2806-6

90000

9 780813 828060